JOHN MUIR

WILDERNESS ESSAYS

Gibbs M. Smith, Inc.
Peregrine Smith Books
Salt Lake City

LITERATURE OF THE AMERICAN WILDERNESS

Introduction Copyright © 1980 by Peregrine Smith, Inc.

Library of Congress Cataloging in Publication Data

Muir, John, 1838-1914.
  Wilderness essays.

    (Literature of the American wilderness)
    1. Natural history—United States—
Addresses, essays, lectures. I. Title.
II. Series.
QH104.M85   917.9'042   80-10844
ISBN 0-87905-072-1

Manufactured in the United States of America.

Cover painting: *Yosemite Valley* (1868) by Albert Bierstadt.

Fourth printing, 1985

Published by Gibbs M. Smith, Inc.
P.O. Box 667, Layton, Utah 84041

JOHN MUIR
WILDERNESS ESSAYS

LITERATURE OF THE AMERICAN WILDERNESS

The Peregrine Smith Literature of the American Wilderness series is designed to make available the outstanding natural history writings of the North American continent from the first settlements and discoveries to the present day. All volumes in the series will feature uniform attractive designs and modest prices to enable all readers to add these beautiful volumes to their library. Outstanding new introductions by noted scholars, poets and wilderness writers will enhance readers' understanding of these classic volumes celebrating the beauty and wonder of the American earth and experience.

Editorial advisers to the series are William Howarth of Princeton University, and Frank Bergon of Vassar College.

Titles in print:

John Muir, *Wilderness Essays*. Edited with an introduction and notes by Frank Buske.

John Muir, *Mountaineering Essays*. Edited with an introduction by Richard F. Fleck.

Theodore Roosevelt, *Wilderness Writings*, edited with an introduction by Paul Schullery.

Henry David Thoreau. *The Natural History Essays*. Edited with an introduction and notes by Robert Sattelmeyer.

Henry David Thoreau, *The Journal*, all fourteen original volumes with an added botanical index to the *Journal*, in a handsome paperback set in a slipcase. Introductions by Walter Harding.

John C. Van Dyke, *The Desert*. Introduction by Richard Shelton.

CONTENTS

INTRODUCTION
"IT IS STILL THE MORNING OF CREATION"

In 1975, during spring quarter at the University of California, Davis, I taught a course for the Division of Environmental Studies called "Nature Writers in Nineteenth-Century America." The final works the students read were by John Muir: *The Mountains of California* and several of the essays included in this volume, "Twenty Hill Hollow," "Wild Wool," and "The Animals of Yosemite."

The next weekend the class and I travelled to Yosemite, which had provided much of the material for *The Mountains of California*, to look at—and try to see—the things that Muir had seen and described. We found a water ouzel's nest, with newly-hatched young, under a footbridge over the Merced River; we walked close to Yosemite Falls and found the place where Muir had once built a cabin; we saw Half Dome and climbers on El Capitan. That night we sat atop Sentinel Dome and listened to the music of the waterfalls as the stars came out; the next morning we had breakfast at Glacier Point and fed the birds and animals that gathered in friendly fashion around us. If we had been less inhibited, we might have exalted, as Muir once did upon returning from a trip to the High Sierras: "It is still the morning of creation, the morning stars are singing together and all the Sons of God shouting for joy!"

Who was this man John Muir whose words, written scores of years before, still had the power to move readers to try to appreciate the wonders of the world as he saw them? My own interest in Muir was rather easily explained: I had grown up on a farm in Wisconsin not far from the region in which Muir, his parents and brothers and sisters, had cleared a homestead in the wilderness. Later, after moving to Alaska, I had travelled through the same areas to which Muir had thrilled on his early visits there and saw the huge ice masses still sculpting the landscape, the glacier mills still grinding the rocks into soil, the mountain fountains in ceaseless flow. The air was still soft and like a poultice; the land and waters still teemed with plant and animal life. Much had changed, but it was still a land John Muir would have recognized and in which he would have felt comfortable.

John Muir, born in Scotland in 1838, crossed the Atlantic in a sailing ship in 1849 with his father Daniel, his brother David, and his sister Sarah. After they had settled in Wisconsin, clearing the land and building a house, the rest of the family joined them: Mrs. Muir, two brothers and a sister.

For the next ten years, John Muir knew a life that would have been familiar to any pioneer on any of America's frontiers. Although he had attended school regularly in Scotland, there was time for only two months of formal education during

INTRODUCTION

the next decade. He was too busy with the breaking plow, chiselling a well through a thick limestone layer (and nearly losing his life when gas collected in the bottom of the excavation), and doing all the other chores normal to making a living under those circumstances. Daniel Muir believed that the good life, for his sons and daughters, consisted of long hours of hard work during the day and learning Bible verses at night. In Scotland, Muir had read whenever possible, as much as he could, and although his father believed that the Bible was the only book human beings could possibly require throughout their journey from earth to heaven, he did relent when he could be persuaded of the strong moral purpose of a work. Milton and Shakespeare were permitted, but Plutarch was forbidden until Muir suggested that the work might contain valuable dietary advice (this at a time when his father was attempting to enforce strict vegetarianism on the family).

And yet life on the frontier did have its rewards. In writing of "that glorious Wisconsin wilderness" in *The Story of My Boyhood and Youth*, Muir spoke of the "sudden plash into pure wildness—baptism in Nature's warm heart," and of the lessons in nature that he and his brothers and sisters learned. These experiences became the solid background for the life he would later choose to lead.

Muir left home in the autumn of 1860 without any clear purpose in mind except to display his inventions at the state fair in Madison. Early in

1861, just short of his twenty-third birthday, he entered the University of Wisconsin after a brief preparatory course to make up for his lack of formal schooling.

At the university, Muir signed up for courses with James Davie Butler, professor of classics, and Ezra Slocum Carr, professor of natural history, both of whom befriended Muir and were to influence him profoundly. In addition, he met Carr's wife, Jean, an intelligent and energetic woman with a strong interest in botany and plant collecting. During their years in Madison, the Carrs had met and become friends with America's most eminent transcendentalist, Ralph Waldo Emerson, on one of his lecture trips to that city. Mrs. Carr read Muir Emerson's poem, "Woodnotes," and introduced him to other works of Emerson and to the writings of Henry David Thoreau. She also introduced him to those of her friends in Madison who shared her interests in nature and the relationship of God and man to it.

Muir's years at the University were an exhilarating and liberating experience; they were also times of difficulty for him. Desperately poor—he received no help from his father—he worked at whatever jobs he could find to try to stay in school, even teaching one winter in a grammar school in a small community near Madison. Also, the Civil War had broken out; he was appalled by the senselessness of war and troubled by his visits to young men from his home neighborhood whom he visited at a camp

INTRODUCTION

near Madison. It must have occured to him that he might be called to serve. After deciding to enter medical school at the University of Michigan, he set out in June of 1863 on a plant-gathering summer's ramble through Indiana and Michigan, finally making his way to Canada.

The ramble lasted for four years. Muir stopped at least twice during this time to work in factories, putting his inventive talents to work to improve production. He might, in fact, have had a successful career in industry. But when a dropped file was propelled into his eye by a moving belt and he nearly lost his sight, Muir decided to devote his life to a study of the natural world.

During his travels, Muir collected plant specimens and kept up a correspondence with Mrs. Carr. In a letter to her, written while recovering from his eye injury, he recalled that he had read an account of the Yosemite Valley the year before, and the two of them discussed in subsequent letters the possibility that one day the Carrs and Muir would, in fact, see Yosemite Valley. In good health again, Muir returned to Wisconsin to say goodbye to his family and shortly thereafter began the journey that took him to California.

It was in September, 1867, that Muir left Indianapolis to begin his 1,000 mile walk to the Gulf of Mexico, a journey that he expected to continue on to the Amazon in South America. He carried in his pack small volumes of the poems of Robert Burns, Milton's *Paradise Lost*, the New Testament, and a journal on the inside cover of which he had written, "John Muir, Earth-Planet, Universe." Professor Butler had urged him years before to keep a journal; he was now ready to heed that advice.

The journal that Muir kept during his walk was his first conscious literary effort, his first attempt to express what he saw and what he felt about what he saw. The South, and the people who lived there, then in the troubled times of the Reconstruction, did not interest him nearly as much as the new plants he was seeing. He put down his skepticism about the importance of man in the universe: "Why should man value himself as more than a small part of the great creation?" True, the universe would be incomplete without man, but it would also be incomplete without even the smallest microscopic creature. Man had arrived rather late on planet earth and would one day take his place among creatures that had once existed and had since returned to dust.

By the time Muir reached Cuba, he had suffered two attacks of fever which left him too weak for strenuous travel. His money was running low; he took passage to California by way of Panama and a connecting ship to San Francisco. Upon arrival there, he started to walk eastward, toward the Sierras, which became for him the "Range of Light." To support himself, he took a job herding sheep—"hoofed locusts" he would call them later—and began his systematic study of the Yosemite

INTRODUCTION

region.

The official version of the origin of Yosemite Valley, promulgated by Dr. Josiah Whitney, California State Geologist, was that some cataclysm had caused the formation of the huge valley; Muir became convinced that glacial action had been responsible for carving it. Learning of Muir's theories, Whitney dismissed Muir as "a mere sheepherder"; but other scientists were more impressed: Joseph Le Conte, professor of geology at the University of California, and John Daniel Runkle, president of the Massachusetts Institute of Technology, who offered him a position teaching at his school. Muir's theories were beginning to gain acceptance.

In 1869, Professor Carr resigned his professorship at the University of Wisconsin and the Carrs moved west to Oakland. Muir's correspondence with Mrs. Carr had continued, but now the friends could exchange visits as well. In addition to sending Muir letters, Mrs. Carr also sent him visitors, the most famous of these being Ralph Waldo Emerson who arrived in the valley in May, 1871. It would be interesting to know what the two men talked about, but neither man ever wrote a detailed account of their conversations. The two became friends; Emerson wanted Muir to come back to the east coast to spread his knowledge about glacial action, but Muir felt that his work in the Yosemite was not yet complete. Muir sent Emerson cedar boughs with blossoms; Emerson sent Muir books. Late in life, Emerson added John Muir's name to his list of "My Men," men who had had the greatest effect on him.

From the beginning of his studies in Yosemite Valley, friends had urged Muir to publish accounts of his findings and theories. It was a period during which the nature essay was gaining in popularity: during the 1860s, collections of Thoreau's essays had come into print and John Burroughs was beginning to write his essays about familiar birds, animals and plants. Muir had sent several accounts of his experiences in Yosemite Valley to the New York *Tribune*, all of which had been published. Mrs. Carr was particularly anxious that Muir write not only articles, but possibly even a book, detailing his glacier theories and observations, though she did have some reservations about the lack of polish in his writing style. Finally, Muir came down from the mountains and began to organize his materials and to put them into articles. The *Overland Monthly* began printing them in April, 1872.

At this stage in his life, Muir certainly did not regard himself as a literary figure of any kind— the fact that he had to so regard himself in later years was painful to him. But he had always been fond of words and the way they could be used, and he had a gift for precise observation and the ability to express his thoughts in vivid imagery. Early in his life he had made the discovery "that the poetry of the Bible, Shakespeare, and Milton was a source of inspiring, exhilarating, uplifting pleasure." The

INTRODUCTION

nature essay was probably as congenial a genre as he could have found.

"Twenty Hill Hollow," which appeared in the *Overland Monthly* in July, 1872, describes a spot Muir had visited during his first summer in California. Muir's piece gives a careful description of the geology of the area, the plants, animals, birds, and weather to be found there, but it is the conclusion that startles. Muir writes that for a visitor, "plain, sky, and mountains ray beauty which you feel. You bathe in these spirit-beams.... Presently you lose consciousness of your own separate existence; you blend with the landscape, and become part and parcel of nature." This state of mystic communion with nature was very like the transparent eyeball that Emerson had described in his important essay, "Nature."

Early in his studies in the Yosemite, Muir formulated his most famous image, the metaphor of mountains as "fountains of men." The process began with one of the simplest of all of nature's manifestations, the snowflake (he called them "snowflowers"). The snowflakes fell, accumulating in huge layers over the land. When the climate warmed, these great snow masses, now compressed into ice as glaciers, began to melt and to move, becoming a powerful sculpting force that carved the mountains into ridges and canyons and becoming, too, a mighty mill that ground the rocks into soils. The water from the melting carried these soils down slope, depositing them in meadows and on flatlands where plants and trees began to grow. Soon animals came to live in these areas, then men. And the process was continuous. Wherever there were glaciers, the world was in a constant state of creation. Snowflowers also provided man with another kind of beauty, however. When conditions were right, the wind picked up these snowflowers, whirled them through the air, and they streamed from the mountain tops in long snow banners, fluttering against a vivid, clear blue sky.

In "A Near View of the High Sierra," Muir presents one of his most artful pieces of writing. He begins with an almost painterly (he made many fine, quick sketches during his travels) description of the Sierras in autumn while returning to the Yosemite Valley. There he meets two painters who want to do some sketching so he guides them to a place where they will find the kinds of scenery they wish to draw. Then he leaves them to make an ascent of Mount Ritter. When he reaches the top he beholds a panorama they will never be able to see or to paint and the contrast between the two painters who are satisfied with the part and with Muir who sees the whole is particularly striking.

The Sierra was filled with many wonders for Muir, not the least of them the wild animals. He had always had a fondness for animals since his boyhood in Scotland and Wisconsin, and though he had to overcome aversions to alligators and rattlesnakes, he came to believe that all living things were brothers and sisters to man (though it is

INTRODUCTION

doubtful that he ever learned to like sheep). He loved the Douglas squirrel and liked to listen to— and retell—bear stories. In fact, Muir liked anything that was wild, and in "Wild Wool" he has no difficulty demonstrating to his satisfaction that wool from wild sheep is far superior to that of domesticated breeds. Anything and anyone living in a direct relationship with nature he could appreciate.

In 1874, Muir began writing a series of letters for the San Francisco *Bulletin*, describing his investigations in the Sierra. When, in 1878, he crossed the Sierras into Nevada and Utah, he sent back to the newspaper descriptions of the landscape and some rather biased accounts of his meetings with the Mormons. One of his most enjoyable experiences was a "Great Storm in Utah," an experience as vivid as the night he clung to a pine tree during a wind storm in the Sierras. All natural phenomena thrilled Muir: on one occasion he was so enthusiastic about an earthquake in the Yosemite that people thought he must be out of his mind!

By 1879, Muir had completed most of the studies he wished to make in Yosemite Valley. At a Sunday School convention there in June of that year, he met Sheldon Jackson, a Presbyterian missionary who had pioneered a mission effort in Alaska. Muir had been thinking about travelling northward; he determined to go to Alaska.

Alaska became the great adventure of Muir's life. There were glaciers everywhere, fronting on rivers and bays of the ocean; the climate was mild and invigorating; the land was even more remote and untouched than that of the Yosemite. Although he had become engaged to be married just before leaving for Alaska, he could not tear himself away from this new land. In October, 1879, Muir and the resident missionary at Fort Wrangel, S. Hall Young, set out with four Indian paddlers—Toyatte, Kadachan, Stickeen John and Sitka Charley—to look for a bay filled with ice. When they reached their destination, Muir and Young became the first white men to explore what later came to be known as Glacier Bay; the largest ice mass in it was later named Muir Glacier.

After Muir's marriage in 1880, his travel and writing activities decreased for the next decade while he engaged in fruit ranching in Martinez, California, to provide for his wife and the two daughters born to them. His major literary effort was to edit *Picturesque California and the Regions West of the Rocky Mountains from Alaska to Mexico*, an over-size, two-volume set for which he got some of the West's most prominent authors and artists to contribute articles, drawings and paintings. Muir wrote seven of the essays, and in his description of the Columbia River basin included material which later appeared separately as "The Forests of Oregon." The piece gave him an opportunity to enthuse about his favorite tree, the sugar pine, together with an account of its discovery by David Douglas, a Scottish botanical

INTRODUCTION

explorer. The book also contains a brief account of Muir and Young's exploration of Glacier Bay, the first time an account of that voyage, written by Muir, had appeared in print.

After 1890, Muir relinquished management of his fruit ranch and turned to writing up some of the notes he had accumulated during his years of exploration, and to the cause of the preservation of America's forests and natural wonders. One of the founders of the Sierra Club in 1891, he became its first president, a post in which he served until his death. Muir's interest in conservation was no recent thing. In February of 1876, the Sacramento *Union* had published his letter, "How Shall We Preserve Our Forests?" His first-hand knowledge of the destructive activity of the sheep in the Sierras made him a passionate advocate of those who would save the beauties of nature from those who would despoil or destroy them.

During the 1890s, Muir's writing took on more and more a purpose he had declared in an earlier letter to Mrs. Carr: "I care to live only to entice people to look at Nature's loveliness." *The Mountains of California*, Muir's first, and considered by many readers his best, book appeared in 1894. The book contained many of the essays that had appeared earlier and that had established Muir's fame as a nature writer, particularly "The Water Ouzel," which is still widely anthologized. Muir was interested not only in describing the landscape but also in showing man's relationship to it. The feeling was explicit that any reader who would travel to the Sierras would be able to realize great benefits from the experience.

His travels in Alaska had so inspired Muir that he always said that he wanted to write a number of books about them, but he did not actually begin to write extensively about the area until after his fourth trip there in 1890 (he would make three more during his lifetime). "The Discovery of Glacier Bay" appeared in *Century Magazine* in June of 1895; by that time thousands of tourists had seen not only the Bay but also Muir Glacier. Tourist steamships had begun including Glacier Bay on their itineraries as early as 1883, and a number of Alaskan guide books had already extolled its beauties; even the Baedeker handbook for the United States recommended Muir Glacier as an outstanding tourist attraction. Muir's account, written sixteen years after that voyage of discovery, recalls vividly that dark and huge journey through the gloomy autumn weather, the hardships and the dangers. It is one of the great travel narratives in American literature.

"The Alaska Trip," made a timely appearance in *Century Magazine*, August, 1897, the very month the gold rush to the Klondike was at a peak of feverish intensity. In the article, Muir frankly extolled the beauties of Alaska and urged tourists to go to see them, claiming that not only will travellers be impressed with the beauty they find there, but also their health will be vastly improved by the bracing, invigorating air. Materials for this article were drawn from the newspaper letters Muir had sent back to the San Francisco *Bulletin* during his trips to Alaska in 1879 and 1880. Muir later claimed that it was he who had made Alaska a popular tourist attraction, and there was considerable merit in the claim.

In an effort to promote greater interest in the national parks and forests, Muir wrote a series of articles for the *Atlantic Monthly*. Yellowstone Park, which he had visited in 1885, was the subject of an article in April, 1898. It was an area that particularly intrigued Muir because of the activity of the bubbling pots of hot mud, the sulphurous cauldrons, and the jetting geysers. The account was later included in Muir's second book, *Our National Parks*, which came out in 1901.

By the early 1900s, Muir had a worldwide reputation as a writer and naturalist. President Theodore Roosevelt went camping with him in the Yosemite Valley; his help was constantly sought in battles involving conservation. After the death of his wife in 1905, Muir undertook some of the travels he'd always dreamed of: to the rain forests of the Amazon and to see the baobab tree in Africa. There was no time for all the things he wanted to do; he did not get to write all the books he planned.

Muir's most popular book, *Stickeen*, the story of crossing Taylor Glacier with S. Hall Young's mongrel dog, became an immediate popular success when it appeared in 1909. John Burroughs complained that when anyone asked Muir to tell the story they always got the whole theory of glacial activity thrown in! *My First Summer in the Sierra*, Muir's journal of his earliest explorations in the Sierra, appeared in 1911. *The Yosemite*, a tourist guidebook to that region, was published in 1912. He told of the years of his growing up in *The Story of My Boyhood and Youth* in 1913. The book that he had waited so long to write, *Travels in Alaska*, lay unfinished on his bedside table when Muir died on Christmas Eve, 1914. In a world, in a universe, which Muir regarded as being in a continuing state of creation, this seems, at least, appropriate.

Muir's reputation as a writer rests not only on the books and articles he published during his lifetime. *Travels in Alaska* was published posthumously and the latter portions of the book show that he had not worked the materials into final form. William Frederic Badè, Muir's literary executor, fashioned three more books from Muir's papers. *A Thousand Mile Walk to the Gulf*, Muir's first journal, on which he had done some editorial work, came out in 1917. *The Cruise of the Corwin*, made up of letters Muir had written to the San Francisco *Bulletin* while aboard the *Corwin* which searched for the *Jeanette*, a ship lost in the ice of the Arctic Ocean, appeared in 1917. *Steep Trails* was made up of articles from various sources, newspaper and magazine, describing Muir's travels in areas of the West other than California; it was published in 1918. In 1938, Linnie Marsh Wolfe edited the materials in the notebooks and on the scraps of papers found after Muir's death, compiling *John of the Mountains: The Unpublished Journals of John Muir*. In addition, Muir's daughters arranged for the appearance of a significant work, *Letters to a Friend, Written to Mrs. Ezra S. Carr, 1866-1879*, in 1915, showing Muir's developing interest in nature and his commitment to a vocation of the study of it.

How shall we assess the importance of Muir as a nature writer? As in the case of Thoreau, Muir's personality frequently intrudes on our evaluation of him as a writer. Muir was a quintessential romantic frontier figure. Unarmed, carrying only a few crusts of bread, a tin cup, a small portion of tea, a notebook and a few scientific instruments, Muir walked into the vastness of the Sierras to search out truths. Single-minded, he did not hesitate to challenge the accepted authorities and their explanations regarding the wilderness he loved: he formulated his own theories and carefully searched out the evidence. America has always loved its rebels, even if it turns out later that they have not discovered the whole truth.

Part of Muir's attractiveness to modern readers is the fact that he was an activist. He not only explored the west and wrote about its beauties—he fought for their preservation. His successes dot the landscape in all the natural features that bear his name: forests, lakes, trails, glaciers. His writings still stimulate readers to try to retrace his footsteps through areas about which he wrote so compellingly.

Muir's prose is clean and direct, because his powers of observation were keen, and he drew his figures of speech, his similes and metaphors from the natural world. He is frequently "adjectivorous," as he admits, but at least part of his problem arises from the vastness of the landscape he is trying to describe. His transcendental brother, Thoreau, had found his correspondences between man and nature at Walden Pond, a physical area which would have filled only a small portion of Yosemite Valley which, in turn, is only a small corner of the Sierras, in turn only a fragment of the American West.

In commenting on Muir's achievement, Norman Foerster, in *Nature in American Literature*, wrote: "Whoever would know the Far West, from Alaska to Mexico, from the coast to the Rockies, must know John Muir.... [he] gave this region to the country—both to those who could not go to see and to those who, having eyes, saw not. That is his foremost achievement."

But Muir did much more than that: in the forests and national parks he helped to preserve, he gave the natural world back to the people of America. Had he never written a single word, that alone would be an impressive monument.

Frank E. Buske
University of Alaska
January, 1980

THE DISCOVERY OF GLACIER BAY BY ITS DISCOVERER

My first visit to the now famous Glacier Bay of Alaska was made toward the end of October, 1879, when young ice was beginning to form in the branch inlets occupied by the glaciers, and the mountains were mantled with fresh snow all the way down from the highest peaks and ridges of the Fairweather Range nearly to the level of the sea.

I had spent most of the season exploring the cañon of the Stickeen River and its glaciers, and a small portion of the interior region beyond the Coast Mountains, on the divide of some of the southerly tributaries of the Yukon and Mackenzie rivers. When I got back to my headquarters at Fort Wrangel, about the beginning of October, it seemed too late for new undertakings in this icy northland. The days were growing short, and winter, with its heavy storms, was drawing nigh, when avalanches would be booming down the long white slopes of the peaks, and all the land would be buried. But, on the other hand, though this white wilderness was new to me, I was familiar with storms, and enjoyed them, knowing well that in right relations with them they are ever kindly. The main inland channels, extending in every direction along the coast, remain open all winter;

The Century Magazine (June, 1895)

and their shores being well forested, it would be easy to keep warm in camp, while in a large canoe abundance of provisions could be carried. I determined, therefore, to go ahead as far north as possible, with or without companions, to see and learn what I could, especially with reference to future work. When I made known my plans to Mr. Young, the Wrangel missionary, he offered to go with me, and with his assistance I procured a good canoe and a crew of Indians, gathered a large stock of provisions, blankets, etc., and on October 14 set forth, eager to welcome whatever wildness might offer, so long as food and firewood should last.

Our crew numbered four: Toyatte, a grand old Stickeen nobleman, who was elected captain, not only because he owned the canoe, but for his skill in woodcraft and seamanship; Kadechan, the son of a Chilcat chief; John, a Stickeen who acted as interpreter; and Sitka Charlie. Mr. Young is one of those fearless and adventurous evangelists who in seeking to save others save themselves, and it was the opportunities the trip might afford to meet the Indians of the different tribes along our route that induced him to join me.

After all our bundles were stowed aboard, and we were about to cast loose from the wharf, Kadechan's mother, a woman of great natural dignity and force of character, came down the steps alongside the canoe, oppressed with anxious fears for the safety of her son. Standing silent for a few moments, she held the missionary with her dark, bodeful eyes, and at length, with great solemnity of speech and gesture, accused him of using undue influence in gaining her son's consent to go on a dangerous voyage among tribes that were unfriendly to the Stickeens. Then, like an ancient sibyl, she foretold a long train of disasters from stormwinds and ice, and in awful majesty of mother-love finished by saying: "If my son comes not back, on you will be his blood, and you shall pay. I say it." Mr. Young tried in vain to calm her fears, promising Heaven's care as well as his own for her precious son, assuring her that he would faithfully share every danger that might assail him, and, if need be, willingly die in his defense. "We shall see whether or not you die," she said as she turned away.

Toyatte also encountered domestic difficulties in getting away. When he stepped into the canoe I noticed a cloud on his grand old face, as if his sad doom, now drawing near, was already beginning to overshadow him. When he took leave of his wife she wept bitterly, saying that the Chilcat chiefs would surely kill him in case he should escape the winter storms. But it was not on this trip that the old hero was to meet his fate, and when we were fairly free in the wilderness these gloomy forebodings vanished, and a gentle breeze pressed us joyfully onward over the shining waters.

We first pursued a westerly course through Sumner Strait, between Kupreanof and Prince of Wales islands; then, turning northward, we sailed up the charming Kiku Strait, through the midst of innumerable picturesque islets, across Prince Frederick Sound, up Chatham Strait, and thence northwestward through Icy Strait and around Glacier Bay. Thence, returning through Icy Strait, we urged our way up the grand Lynn Canal to the Davidson Glacier and Chilcat, and returned to Wrangel along the coast of the mainland, visiting the icy Sum Dum Bay and the Le Conte Glacier on our route. Thus we made a journey more than eight hundred miles long; and though hardships were encountered, and a few dangers, the wild wonderland made compensation beyond our most extravagant hopes.

The first stages of our journey were mostly enjoyment. The weather was about half bright, and we glided along the green and yellow shores in comfort, the lovely islands passing in harmonious succession, like ideas in a fine poem. The rain did not hinder us, but when the wind was too wild we stayed in camp, the Indians usually improving such storm times in deer-hunting, while I examined the rocks and woods. Most of our camps were made in nooks that were charmingly embowered, and fringed with bushes and late flowers. After supper we sat long around the fire, listening to the stories of the Indians about the wild animals they were acquainted with, their hunting adventures, wars, traditions, religion, and customs. Every Indian party we met we interviewed, and every village we came to we visited.

DISCOVERY OF GLACIER BAY

Thus passed our days and nights until we reached the west coast of Admiralty Island, intending to make a straight course thence up Lynn Canal, when we learned from a party of traveling Hoonas that the Chilcats had been drinking and quarreling, that Kadechan's father had been shot, and that we could not go safely into their country before these whisky quarrels were settled. My Indians evidently believed this news, and dreaded the consequences; therefore I decided to turn to the westward through Icy Strait, and to go in search of the wonderful ice-mountains to which Sitka Charlie, the youngest of my crew, had frequently referred. Having noticed my interest in glaciers, he told me that when he was a boy he had gone with his father to hunt seals in a large bay full of ice, and that he thought he could find it if I cared to have him try. I was rejoiced to find all the crew now willing to go on this adventure, judging, perhaps, that ice-mountains under the present circumstances might prove less dangerous than Chilcats.

On the 24th, about noon, as we came near a small island in Icy Strait, Charlie said that we must procure some dry wood there, for in the ice-mountain country which we were now approaching not a single tree of any kind could be found. This seemed strange news to the rest of the crew, and I had to make haste to end an angry dispute that was rising by ordering as much wood to be taken aboard as we could carry. Then we set sail direct for the ice-country, holding a north-westerly course until long after dark, when we reached a small inlet that sets in near the mouth of Glacier Bay, on the west side. Here we made a cold camp on a desolate snow-covered beach in stormy sleet and darkness. At daybreak I looked eagerly in every direction to learn what kind of place we were in; but gloomy rainclouds covered the mountains, and I could see nothing that could give me a clue, while Vancouver's chart, hitherto a faithful guide, here failed us altogether. Nevertheless, we made haste to be off; and fortunately, just as we were leaving the shore, a faint smoke was seen across the inlet, toward which Charlie, who now seemed lost, gladly steered. Our sudden appearance so early that gray morning had evidently alarmed our neighbors, for as soon as we were within hailing distance an Indian with his face blackened fired a shot over our heads, and in a blunt, bellowing voice roared, "Who are you?" Our interpreter shouted, "Friends and the Fort Wrangel missionary." Then men, women, and children swarmed out of the hut, and awaited our approach on the beach. One of the hunters having brought his gun with him, Kadechan sternly rebuked him, asking with superb indignation whether he was not ashamed to bring a gun in his hand to meet a missionary. Friendly relations, however, were speedily established, and as a cold rain was falling, they invited us into their hut. It seemed small for two persons; nevertheless, twenty-one managed to find shelter in it about a smoky fire. Our hosts proved to

be Hoona seal-hunters laying in their winter stores
of meat and skins. The packed hut was passably
well ventilated, but its oily, meaty smells were not
the same to our noses as those of the briny, sprucy
nooks we were accustomed to, and the circle of
black eyes peering at us through a fog of reek and
smoke made a novel picture. We were glad, however, to get within reach of information, and of
course asked many questions concerning the ice-
mountains and the strange bay, to most of which
our inquisitive Hoona friends replied with counter-
questions as to our object in coming to such a
place, especially so late in the year. They had
heard of Mr. Young and his work at Fort Wrangel,
but could not understand what a missionary could
be doing in such a place as this. Was he going to
preach to seals and gulls, they asked, or to the
ice-mountains? and could they take his word? Then
John explained that only the friend of the missionary was seeking ice-mountains; that Mr.
Young had already preached many good words in
the villages we had visited on our way, in their own
among the rest; that our hearts were good; and that
every Indian was our friend. Then we gave them a
little rice, sugar, tea, and tobacco, after which they
began to gain confidence and to speak freely. They
told us that the main bay was called by them
Sit-a-da-kay, or Ice Bay; that there were many
large ice-mountains in it, but no gold mines; and
that the ice-mountain they knew best was at the
head of the bay, where most seals were found.

WILDERNESS ESSAYS

Notwithstanding the rain, I was anxious that we should push and grope our way beneath the clouds as best we could, in case worse weather should come; but Charlie was ill at ease, and wanted one of the seal-hunters to go with us, for the place was much changed. I promised to pay well for a guide, and in order to lighten the canoe proposed to leave most of our heavy stores with our friends until our return. After a long consultation one of them consented to go. His wife got ready his blanket and a piece of cedar matting for his bed, and some provisions—mostly dried salmon, and seal sausage made of strips of lean meat plaited around a core of fat. She followed us to the beach, and just as we were pushing off said with a pretty smile: "It is my husband that you are taking away. See that you bring him back." We got under way about 10 A.M. The wind was in our favor, but a cold rain pelted us, and we could see but little of the dreary, treeless wilderness which we had now fairly entered. The bitter blast, however, gave us good speed; our bedraggled canoe rose and fell on the icy waves, solemnly bowing to them, and mimicking the gestures of a big ship. Our course was northwestward, up the southwest side of the bay, near the shore of what seemed to be the mainland, some smooth marble islands being on our right. About noon we discovered the first of the great glaciers—the one I afterward named for Geikie, the noted Scotch geologist. Its lofty blue cliffs, looming up through the draggled skirts of the clouds, gave a tremendous impression of savage power, while the roar of the new-born icebergs thickened and emphasized the general roar of the storm. An hour and a half beyond the Geikie Glacier we ran into a slight harbor where the shore is low, dragged the canoe beyond the reach of drifting icebergs, and, much against my desire to push ahead, encamped, the guide insisting that the big ice-mountain at the head of the bay could not be reached before dark, that the landing there was dangerous even in daylight, and that this was the only safe harbor on the way to it. While camp was being made I strolled along the shore to examine the rocks and the fossil timber that abound here. All the rocks are freshly glaciated even below the sea-level, nor have the waves as yet worn off the surface polish, much less the heavy scratches and grooves and lines of glacial contour.

The next day being Sunday, the minister wished to stay in camp; and so, on account of the weather, did the Indians. I therefore set out on an excursion, and spent the day alone on the mountain slopes above the camp, and to the north of it, to see what I might learn. Pushing on through rain and mud and sludgy snow, crossing many brown, boulder-choked torrents, wading, jumping, wallowing in snow to my shoulders, I had a desperately hard and dangerous time. After crouching cramped and benumbed in the canoe, poulticed in wet clothes and blankets night and day, my limbs had been long asleep. This day they were awake, and in the

hour of trial proved that they had not lost the
cunning learned on many a mountain peak of the
high Sierra. I reached a height of 1500 feet, on
the ridge that bounds the second of the great
glaciers on the south. All the landscape was
smothered in clouds, and I began to fear that I
had climbed in vain, as far as wide views were
concerned. But at length the clouds lifted a little,
and beneath their gray fringes I saw the berg-
filled expanse of the bay, and the feet of the
mountains that stand about it, and the imposing
fronts of five of the huge glaciers, the nearest
being immediately beneath me. This was my first
general view of Glacier Bay, a solitude of ice and
snow and new-born rocks, dim, dreary, mysterious.
I held the ground I had so dearly won for an hour or
two, sheltering myself as best I could from the
blast, while with benumbed fingers I sketched what
I could see of the landscape, and wrote a few lines
in my note-book. Then I breasted the snow again,
crossed the muffled, shifting avalanche tali, forded
the torrents in safety, and reached camp about
dark, wet and weary, but rich in a notable
experience.

While I was getting some coffee, Mr. Young told
me that the Indians were discouraged, and had
been talking about turning back, fearing that I
would be lost, or that in some way the expedition
would come to grief if I persisted in going farther.
They had been asking him what possible motive I
could have in climbing dangerous mountains when

blinding storms were blowing; and when he replied that I was only seeking knowledge, Toyatte said, "Muir must be a witch to seek knowledge in such a place as this, and in such miserable weather." After supper, crouching about a dull fire of fossil wood, they became still more doleful, and talked in tones that accorded well with the growling torrents about us, and with the wind and rain among the rocks, telling sad old stories of crushed canoes and drowned Indians, and of hunters lost and frozen in snow-storms. Toyatte, dreading the treeless, forlorn appearance of the region, said that his heart was not strong, and that he feared his canoe, on the safety of which our lives depended, might be entering a skookum-house (jail) of ice, from which there might be no escape; while the Hoona guide said bluntly that if I was so fond of danger, and meant to go close up to the noses of the ice-mountains, he would not consent to go any farther: for we should all be lost, as many of his tribe had been, by the sudden rising of bergs from the bottom. They seemed to be losing heart with every howl of the storm, and fearing that they might fail me now that I was in the midst of so grand a congregation of glaciers, which possibly I might not see again, I made haste to reassure them, telling them that for ten years I had wandered along among mountains and storms, and that good luck always followed me; that with me, therefore, they need fear nothing; that the storm would soon cease, and the sun would shine; and that

WILDERNESS ESSAYS

Heaven cared for us, and guided us all the time, whether we knew it or not: but that only brave men had a right to look for Heaven's care, therefore all childish fear must be put away. This little speech did good. Kadechan, with some show of enthusiasm, said he liked to travel with good-luck people; and dignified old Toyatte declared that now his heart was strong again, and he would venture on with me as far as I liked, for my "wawa" was "delait" (my talk was very good). The old warrior even became a little sentimental, and said that if the canoe were crushed he would not greatly care, because on the way to the other world he would have pleasant companions.

Next morning it was still raining and snowing, but the wind was from the south, and swept us bravely forward, while the bergs were cleared from our course. In about an hour we reached the second of the big glaciers, which I afterward named for Hugh Miller. We rowed up its fiord, and landed to make a slight examination of its grand frontal wall. The berg-producing portion we found to be about a mile and a half wide. It presents an imposing array of jagged spires and pyramids, and flat-topped towers and battlements, of many shades of blue, from pale, shimmering, limpid tones in the crevasses and hollows, to the most startling, chilling, almost shrieking vitriol blue on the plain mural spaces from which bergs had just been discharged. Back from the front for a few miles the surface is rendered inaccessible by a series of wide, weathered crevasses, with the spaces between them rising like steps, as if the entire mass of this portion of the glacier had sunk in successive sections as it reached deep water, and the sea had found its way beneath it. Beyond this the glacier extends indefinitely in a gently rising prairie-like expanse, and branches among the slopes and cañons of the Fairweather Range.

From here a run of two hours brought us to the head of the bay, and to the mouth of the northwest fiord, at the head of which lie the Hoona sealing-grounds, and the great glacier now called the Pacific, and another called the Hoona. The fiord is about five miles long, and is two miles wide at the mouth. Here the Hoona guide had a store of dry wood, which we took aboard. Then, setting sail, we were driven wildly up the fiord, as if the storm-wind were saying: "Go, then, if you will, into my ice chamber; but you shall stay until I am ready to let you out." All this time sleety rain was falling on the bay, and snow on the mountains; but soon after we landed the sky began to open. The camp was made on a rocky bench near the front of the Pacific Glacier, and the canoe was carried beyond reach of the bergs and berg-waves. The bergs were now crowded in a dense pack against the ice-wall, as if the storm-wind had determined to make the glacier take back her crystal offspring and keep them at home.

While camp affairs were being attended to, I set out to climb a mountain for comprehensive views; and before I had reached a height of a thousand feet the rain ceased, and the clouds began to rise from the lower altitudes, slowly lifting their white skirts, and lingering in majestic, wing-shaped masses about the mountains that rise out of the broad, icy sea. These were the highest and whitest of all the white mountains, and the greatest of all the glaciers I had yet seen. Climbing higher for a still broader outlook, I made notes and sketched, improving the precious time while sunshine streamed through the luminous fringes of the clouds, and fell on the green waters of the fiord, the glittering bergs, the crystal bluffs of the two vast glaciers, the intensely white, far-spreading fields of ice, and the ineffably chaste and spiritual heights of the Fairweather Range, which were now hidden, now partly revealed, the whole making a picture of icy wildness unspeakably pure and sublime.

Looking southward, a broad ice-sheet was seen extending in a gently undulating plain from the Pacific Fiord in the foreground to the horizon, dotted and ridged here and there with mountains which were as white as the snow-covered ice in which they were half, or more than half, submerged. Several of the great glaciers flow from this one grand fountain. It is an instructive example of a general glacier covering the hills and dales of a country that is not yet ready to be brought to the light of day—not only covering, but creating, a landscape with all the features it is destined to have when, in the fullness of time, the fashioning ice-sheet shall be lifted by the sun, and the land shall become warm and fruitful. The view to the westward is bounded and almost filled by the glorious Fairweather Mountains, the highest of them springing aloft in sublime beauty to a height of nearly 16,000 feet, while from base to summit every peak and spire and dividing ridge of all the mighty host was of a spotless, solid white, as if painted. It would seem that snow could never be made to lie on the steepest slopes and precipices unless plastered on when wet, and then frozen. But this snow could not have been wet. It must have been fixed by being driven and set in small particles like the storm-dust of drifts, which, when in this condition, is fixed not only on sheer cliffs, but in massive overcurling cornices. Along the base of this majestic range sweeps the Pacific Glacier, fed by innumerable cascading tributaries, and discharging into the head of the fiord by two mouths, each nearly a mile wide. This is the largest of all the Glacier Bay glaciers that are at all river-like, the trunk of the larger Muir Glacier being more like a lake than a river. After the continuous rainy or snowy weather which we had had since leaving Wrangel, the clear weather was most welcome. Dancing down the mountain to camp, my mind glowing like the sun-beaten glaciers, I found the Indians seated around a good fire, entirely happy now that the farthest point of the journey had been reached. How keenly bright were the stars that night in the frosty sky, and how impressive was the thunder of the icebergs, rolling, swelling, reverberating through the solemn stillness! I was too happy to sleep.

About daylight next morning we crossed the fiord, and landed on the south side of the island that divides the front wall of the Pacific Glacier. The whiskered faces of seals dotted the water between the bergs, and I could not prevent John and Charlie and Kadechan from shooting at them. Fortunately, they were not skilled in this kind of hunting, and few, if any, were hurt. Leaving the Indians in charge of the canoe, I climbed the island, and gained a good general view of the glacier. At one favorable place I descended about fifty feet below the side of the glacier, where its denuding, fashioning action was clearly shown. Pushing back from here, I found the surface crevassed and sunken in steps, like the Hugh Miller Glacier, as if it were being undermined by the action of the tide-waters. For a distance of fifteen or twenty miles the river-like ice-flood is nearly level, and when it recedes the ocean water will follow it, and thus form a long extension of the fiord, with features essentially the same as those now extending into the continent farther south, where many great glaciers once poured into the sea, though scarce a vestige of them now exists. Thus the domain of the sea has been, and is being, extended in these ice-sculptured lands, and the scenery of the shores is enriched. The dividing island is about a thousand feet high, and is hard beset by the glacier, which still crushes heavily against and around it. A short time ago its summit was at least two thousand feet below the surface of the over-sweeping ice; now three hundred feet of the top is free, and under present climatic conditions it will soon be wholly free from the ice, and will take its place as a glacier-polished island in the middle of the fiord, like a thousand others in this magnificent archipelago. Emerging from its icy sepulcher, it gives a most telling illustration of the birth of a marked feature of a landscape. In this instance it is not the mountain, but the glacier, that is in labor, and the mountain itself is being brought forth.

The Hoona Glacier enters the fiord on the south side, a short distance below the Pacific, displaying a broad and far-reaching expanse, over which many of the lofty peaks of the Fairweather Range are seen; but the front wall, thrust into the fiord, is not nearly so interesting as that of the Pacific, and I did not observe any bergs discharged from it.

After we had seen the unveiling of the majestic peaks and glaciers that evening, and their baptism in the down-pouring sunbeams, it was inconceivable that nature could have anything finer to show us. Nevertheless, compared with what was coming the next morning, all that was as nothing. As far as we could see, the lovely dawn gave no promise of anything uncommon. Its most impressive features were the frosty clearness of the sky, and a deep, brooding calm, made all the more striking by the intermittent thunder of the bergs. The sunrise we did not see at all, for we were beneath the shadows of the fiord cliffs; but in the midst of our studies we were startled by the sudden appearance of a red light burning with a strange, unearthly splendor on the topmost peak of the Fairweather Mountains. Instead of vanishing as suddenly as it had appeared, it spread and spread until the whole range down to the level of the glaciers was filled with the celestial fire. In color it was at first a vivid crimson, with a thick, furred appearance, as fine as the alpenglow, yet indescribably rich and deep—not in the least like a garment or mere external flush or bloom through which one might expect to see the rocks or snow, but every mountain apparently glowing from the heart like molten metal fresh from a furnace. Beneath the frosty shadows of the fiord we stood hushed and awe-stricken, gazing at the holy vision; and had we seen the heavens open and God made manifest, our attention could not have been more tremendously strained. When the highest peak began to burn, it did not seem to be steeped in sunshine, however glorious, but rather as if it had been thrust into the body of the sun itself. Then the supernal fire slowly descending, with a sharp line of demarkation separating it from the cold, shaded region beneath, peak after peak, with their spires and ridges and cascading glaciers, caught the heavenly glow, until all the mighty host stood transfigured, hushed, and thoughtful, as if awaiting the coming of the Lord. The white, rayless light of the morning, seen when I was alone amid the silent peaks of the Sierra, had always seemed to me the most telling of the terrestrial manifestations of God. But here the mountains themselves were made divine, and declared his glory in terms still more impressive. How long we gazed I never knew. The glorious vision passed away in a gradual, fading change through a thousand tones of color to pale yellow and white, and then the work of the ice-world went on again in every-day beauty. The green waters of the fiord were filled with sun-spangles; with the upspringing breeze the fleet of icebergs set forth on their voyages; and on the innumerable mirrors and prisms of these bergs, and on those of the shattered crystal walls of the glaciers, common white light and rainbow light began to glow, while the mountains, changing to stone, put on their frosty jewelry, and loomed again in the thin azure in serene terrestrial majesty. We turned and sailed away, joining the outgoing bergs, while "Gloria in excelsis" still seemed to be sounding over all the white landscape, and our burning hearts were ready for any fate, feeling that whatever the future might have in store, the treasures we had gained would enrich our lives forever.

When we arrived at the mouth of the fiord, and rounded the massive granite headland that stands guard at the entrance on the north side, another large glacier, now named the Reid, was discovered at the head of one of the northern branches of the bay. Pushing ahead into this new fiord, we found that it was not only packed with bergs, but that the spaces between the bergs were crusted with new ice, compelling us to turn back while we were yet several miles from the discharging frontal wall. But though we were not then allowed to set foot on this magnificent glacier, we obtained a fine view of it, and I made the Indians cease rowing while I sketched its principal features. Thence, after steering northeastward a few miles, we discovered still another great glacier, now named the Carroll. But the fiord into which this glacier flows was, like the last, utterly inaccessible on account of ice, and we had to be content with a general view and a sketch of it, gained as we rowed slowly past at a distance of three or four miles. The mountains back of it and on each side of its inlet are sculptured in a singularly rich and striking style of architecture, in which subordinate peaks and gables appear in wonderful profusion, and an imposing conical mountain with a wide, smooth base stands out in the main current of the glacier, a mile or two back from the great ice-wall.

We now turned southward down the eastern shore of the bay, and in an hour or two discovered a large glacier of the second class, at the head of a comparatively short fiord that winter had not yet closed. Here we landed, and climbed across a mile or so of rough boulder-beds, and back upon the wildly broken receding snout of the glacier, which, though it descends to the level of the sea, no longer sends off bergs. Many large masses were detached from the wasting snout by irregular melting, and were buried beneath the mud, sand, gravel, and boulders of the terminal moraine. Thus protected, these fossil icebergs remain unmelted for many years, some of them for a century or more, as shown by the age of trees growing above them, though there are no trees here as yet. At length melting, a pit with sloping sides is formed by the falling of the overlying moraine material into the space at first occupied by the buried ice. In this way are formed the curious depressions in drift-covered regions called kettles, or sinks. On these decaying glaciers we may also find many interesting lessons on the formation of boulders and boulder-beds, which in all glaciated countries exert a marked influence on scenery, health, and fruitfulness.

Three or four miles farther down the bay we came to another fiord, up which we sailed in quest of more glaciers, discovering one in each of the two branches into which the fiord divides. Neither of these glaciers quite reaches tide-water. Notwithstanding their great size and the apparent fruitfulness of their fountains, they are in the first stage of decadence, the waste from melting and evaporation being greater now than the supply of new ice from the snow. We reached the one in the north branch after a comfortable scramble, and climbed over its huge, wrinkled brow, from the top of which we gained a good view of the trunk and some of the tributaries, and also of the sublime gray cliffs that tower on each hand above the ice.

Then we sailed up the south branch of the inlet, but failed to reach the glacier there, on account of a thin sheet of new ice. With the tent-poles we broke a lane for the canoe for a little distance; but it was slow, hard work, and we soon saw that we could not reach the glacier before dark. Nevertheless, we gained a fair view of it as it came sweeping down through its gigantic gateway of massive Yosemite rocks three and four thousand feet high. Here we lingered until sundown, gazing and sketching; then we turned back, and encamped on a bed of cobble-stones between the forks of the fiord.

Our fire was made of fossil wood gathered on the beach. This wood is found scattered or in wave-washed windrows all about the bay where the shores are low enough for it to rest. It also occurs in abundance in many of the ravines and gorges, and in roughly stratified beds of moraine material, some of which are more than a thousand feet in thickness. The bed-rocks on which these deposits rest are scored and polished by glacial action, like all the rocks hereabouts up to at least three thousand feet above the sea. The timber is mostly in the form of broken trunks of the Merten, Paton, and Menzies spruce, the largest sections being twenty to thirty feet long, and from one to three feet in diameter, some of them, with the bark on, sound and tough. It appears, therefore, that these shores were, a century or so ago, as generously forested as those of the adjacent bays and inlets are to-day; though, strange to say, not one tree is left standing, with the exception of a few on mountain-tops near the mouth of the bay and on the east side of the Muir Glacier. How this disforestment was effected I have not space to tell here. I will only say that all I have seen goes to show that the moraine soil on which the forests were growing was held in place on the steep mountain slopes by the grand trunk glacier that recently filled the entire bay as its channel, and that when it melted the soil and forests were sloughed off together.

As we sat by the camp-fire the brightness of the sky brought on a long talk with the Indians about the stars; and their eager, childlike attention was refreshing to see as compared with the decent, deathlike apathy of weary civilized people, in whom natural curiosity has been quenched in toil and care and poor, shallow comfort.

After sleeping a few hours, I stole quietly out of the camp, and climbed the mountain that stands guard between the two glaciers. The ground was frozen, making the climbing difficult in the steep-est places; but the views over the icy bay, sparkling beneath the glorious effulgence of the sky, were enchanting. It seemed then a sad thing that any part of so precious a night had been lost in sleep. The starlight was so full that I distinctly saw not only the bay with its multitude of glittering bergs, but most of the lower portions of the glaciers, lying pale and spirit-like amid the huge silent mountains.

The nearest glacier in particular was so distinct that it seemed to be glowing with light that came from within itself. Not even in dark nights have I ever found any difficulty in seeing large glaciers; but on this mountain-top, amid so much ice, in the heart of so clear and frosty a night, everything was luminous, and I seemed to be poised in a vast hollow between two skies of equal brightness. How strong I felt after my exhilarating scramble, and how glad I was that my good angel had called me before the glorious night succeeding so glorious a morning had been spent!

I got back to camp in time for an early breakfast, and by daylight we had everything packed and were again under way. The fiord was frozen nearly to its mouth, and though the ice was so thin that it gave us but little trouble in breaking a way, yet it showed us that the season for exploration in these waters was well-nigh over. We were in danger of being imprisoned in a jam of icebergs, for the water-spaces between them freeze rapidly, binding the floes into one mass. Across such floes it would be almost impossible to drag a canoe, however industriously we might ply the ax, as our Hoona guide took great pains to warn us. I would have kept straight down the bay from here, but the guide had to be taken home, and the provisions we left at the bark hut had to be got on board. We therefore crossed over to our Sunday storm-camp, cautiously boring a way through the bergs. We found the shore lavishly adorned with a fresh arrival of assorted bergs that had been left stranded at high tide. They were arranged in a broad, curving row, looking intensely clear and pure on the gray sand, and, with the sunbeams pouring through them, suggested the jewel-paved streets of the New Jerusalem.

On our way down the coast, after examining the front of the beautiful Geikie Glacier, we obtained our first broad view of the Muir Glacier, the last of all the grand company to be seen, the stormy weather having hidden it when we first entered the bay. It was now perfectly clear, and the spacious, prairie-like glacier, with its many tributaries extending far back into the snowy recesses of the mountains, made a magnificent display of its wealth, and I was strongly tempted to go and explore it at all hazards. But winter had come, and the freezing of its fiord was an insurmountable obstacle. I had, therefore, to be content for the present with sketching and studying its main features at a distance. When we arrived at the Hoona hunting-camp, the men, women, and children came swarming out to welcome us. In the neighborhood of this camp I carefully noted the lines of demarkation between the forested and disforested regions. Several mountains here are only in part disforested, and the lines separating the bare and the forested portions are well defined. The soil, as well as the trees, had slid off the steep slopes, leaving the edges of the woods raw-looking and rugged.

At the mouth of the bay a series of moraine islands shows that the trunk glacier that occupied the bay halted here for some time, and deposited this island material as a terminal moraine; that more of the bay was not filled in shows that, after lingering here, it receded comparatively fast. All the level portions of trunks of glaciers occupying ocean fiords, instead of melting back gradually in times of general shrinking and recession, as inland glaciers with sloping channels do, melt almost uniformly over all the surface until they become thin enough to float. Then, of course, with each rise and fall of the tide the seawater, with a temperature usually considerably above the freezing-point, rushes in and out beneath them, causing rapid waste of the nether surface, while the upper is being wasted by the weather, until at length the fiord portions of these great glaciers become comparatively thin and weak, and are broken up, and vanish almost simultaneously from the mouths of their fiords to the heads of them.

Glacier Bay is undoubtedly young as yet. Vancouver's chart, made only a century ago, shows no trace of it, though found admirably faithful in general. It seems probable, therefore, that even then the entire bay was occupied by a glacier of which all those described above, great though they are, were only tributaries. Nearly as great a change has taken place in Sum Dum Bay since Vancouver's visit, the main trunk glacier there having receded from eighteen to twenty-five miles from the line marked on his chart.

The next season (1880), on September 1, I again entered Glacier Bay, and steered direct for the Muir Glacier. I was anxious to make my main camp as near the ice-wall as possible, to watch the discharge of the bergs. Toyatte, the grandest Indian I ever knew, had been killed soon after our return to Fort Wrangel; and my new captain, Tyeen, was inclined to keep at a safe distance from the "big ice-mountain," the threatening cliffs of which rose to a height of 300 feet above the water. After a good deal of urging he ventured within half a mile of them, on the east side of the fiord, where with Mr. Young I went ashore to seek a camp-ground on the moraine, leaving the Indians in the canoe. In a few minutes after we landed a huge berg sprung aloft with tremendous commotion, and the frightened Indians incontinently fled, plying their paddles in the tossing waves with admirable energy until they reached a safe shelter around the south end of the moraine, a mile down the inlet. I found a good place for a camp in a slight hollow where a few spruce stumps afforded abundance of firewood. But all efforts to get Tyeen out of his harbor failed. Nobody knew, he said, how far the ice-mountain could dash waves up the beach, and his canoe would be broken. Therefore I had my bedding and some provision carried to a high camp, and enjoyed the wildness alone.

Next morning at daybreak I pushed eagerly back over the snout and along the eastern margin of the glacier, to see as much as possible of the upper fountain region. About five miles back from the front I climbed a mountain 2500 feet high, from the flowery summit of which, the day being clear, the vast glacier and all of its principal branches were displayed in one magnificent view. Instead of a stream of ice winding down a mountain-walled valley, like the largest of the Swiss glaciers, the Muir is a broad, gently undulating prairie surrounded by innumerable icy mountains, from the far, shadowy depths of which flow the many tributary glaciers that form the great central trunk. There are seven large tributaries, from two to six miles wide where they enter the trunk, and from ten to twenty miles long, each of them fed by many secondary tributaries; so that the whole number of branches, great and small, pouring from the mountain fountains must number upward of two hundred, not counting the smallest. The area drained by this one grand glacier can hardly be less than 1000 square miles, and it probably contains as much ice as all the 1100 Swiss glaciers combined. The length of the glacier from the frontal wall back to the head of the farthest fountain is estimated at fifty miles, and the width of the main trunk just below the confluence of the large tributaries is about twenty-five miles. Though apparently as motionless as the mountains, it flows on forever, the speed varying in every part with the seasons, but mostly with the depth of the current, and the declivity, smoothness, and directness of the different portions of the basin. The flow of the central cascading portion near the front, as recently determined by Professor Reid, is at the rate of from two and a half to five inches an hour or from five to ten feet a day. A strip of the main trunk about a mile in width, extending along the eastern margin about fourteen miles to a large lake filled with bergs, has but little motion, and is so little broken by crevasses that one hundred horsemen might ride abreast over it without encountering much difficulty.

But far the greater portion of the vast expanse is torn and crumpled into a bewildering network of hummocky ridges and blades, separated by yawning gulfs and crevasses, so that the explorer, crossing the glacier from shore to shore, must always have a hard time. Here and there in the heart of the icy wilderness are spacious hollows containing beautiful lakes, fed by bands of quick-glancing streams that flow without friction in blue crystal channels, making most delightful melody, singing and ringing in silvery tones of peculiar sweetness, sun-filled crystals being the only flowers on their banks. Few, however, will be likely to enjoy them. Fortunately, to most travelers the thundering ice-wall, while comfortably accessible, is also by far the most interesting portion of the glacier.

The mountains about the great glacier were also seen from this standpoint in exceedingly grand and telling views, peaked and spired in endless variety of forms, and ranged and grouped in glorious array. Along the valleys of the main tributaries to the northwestward I saw far into their shadowy depths, one noble peak appearing beyond the other in its snowy robes in long, fading perspective. One of the most remarkable, fashioned like a superb crown with delicately fluted sides, stands in the middle of the second main tributary, counting from right to left. To the westward the majestic Fairweather Range lifted its peaks and glaciers into the blue sky in all its glory. Mount Fairweather, though not the highest, is by far the noblest of all the sky-dwelling company, the most majestic in port and architecture of all the mountains I have ever seen. It is a mountain of mountains. La Pérouse, at the south end of the range, is also a magnificent mountain, symmetrically peaked and sculptured, and wears its robes of snow and glaciers in noble style. Lituya, as seen from here, is an immense double tower, severely plain and massive. Crillon, though the loftiest of all (being nearly 16,000 feet high), presents no well-marked features. Its ponderous glaciers have ground it away into long, curling ridges until, from this point of view, it resembles a huge twisted shell. The lower summits about the Muir Glacier, like this one, the first that I climbed, are richly adorned and enlivened with beautiful flowers, though they make but a faint show in a general view. Lines and flashes of bright green appear on the lower slopes as one approaches them from the glacier, and a fainter green tinge may be noticed on the subordinate summits at a height of 2000 or 3000 feet. The lower are made mostly by alder bushes, and the topmost by a lavish profusion of flowering plants, chiefly cassiope, vaccinium, pyrola, erigeron, gentiana, campanula, anemone, larkspur, and columbine, with a few grasses and ferns. Of these cassiope is at once the commenest and the most beautiful and influential. In some places its delicate stems make mattresses on the mountain-tops two feet thick over several acres, while the bloom is so abundant that a single handful plucked at random will contain hundreds of its pale pink bells. The very thought of this, my first Alaskan glacier garden, is an exhilaration. Though it is 2500 feet high, the glacier flowed over its ground as a river flows over a boulder; and since it emerged from the icy sea as from a sepulcher it has been sorely beaten with storms; but from all those deadly, crushing, bitter experiences comes this delicate life and beauty, to teach us that what we in our faithless ignorance and fear call destruction is creation.

As I lingered here night was approaching, so I reluctantly scrambled down out of my blessed garden to the glacier, and returned to my lonely camp, and, getting some coffee and bread, again went up the moraine close to the end of the great ice-wall. The front of the glacier is about three miles wide, but the sheer middle, berg-producing portion that stretches across the inlet from side to side, like a huge green-and-blue barrier, is only about two miles wide, and its height above the water is from 250 to 300 feet. But soundings made by Captain Carroll show that 720 feet of the wall is below the surface, while a third unmeasured portion is buried beneath the moraine detritus that is constantly deposited at the foot of it. Therefore, were the water and rocky detritus cleared away, a sheer precipice of ice would be presented nearly two miles long and more than a thousand feet high. Seen from a distance, as you come up the fiord, it seems comparatively regular in form; but it is far otherwise: bold, jagged capes jut forward into the fiord, alternating with deep reëntering angles and sharp, craggy hollows with plain bastions, while the top is roughened with innumerable spires and pyramids and sharp, hacked blades leaning and toppling, or cutting straight into the sky.

The number of bergs given off varies somewhat with the weather and the tides, the average being about one every five or six minutes, counting only those large enough to thunder loudly, and make themselves heard at a distance of two or three miles. The very largest, however, may, under favorable conditions, be heard ten miles, or even farther. When a large mass sinks from the upper fissured portion of the wall, there is first a keen, piercing crash, then a deep, deliberate, prolonged, thundering roar, which slowly subsides into a low, muttering growl, followed by numerous smaller, grating, clashing sounds from the agitated bergs that dance in the waves about the newcomer as if in welcome; and these again are followed by the swash and roar of the waves that are raised and hurled against the moraines. But the largest and most beautiful of the bergs, instead of thus falling from the upper weathered portion of the wall, rise from the submerged portion with a still grander commotion, springing with tremendous voice and gestures nearly to the top of the wall, tons of water streaming like hair down their sides, plunging and rising again and again before they finally settle in perfect poise, free at last, after having formed part of a slow-crawling glacier for centuries. And as we contemplate their history, as we see them sailing past in their charming crystal beauty, how wonderful it seems that ice formed from pressed snow on the far-off mountains two or three hundred years ago should still be pure and lovely in color, after all its travel and toil in the rough mountain quarries in grinding and fashioning the face of the coming landscape! When the sunshine is sifting through the midst of this multitude of icebergs, and through the jets of radiant spray ever plashing from the blows of the falling and rising bergs, the effect is indescribably glorious. Glorious, too, are the nights along these crystal cliffs when the moon and stars are shining. Then the ice-thunder seems far louder than by day, and the projecting buttresses seem higher, as they stand forward in the pale light, relieved by the gloomy hollows, while the new bergs are dimly seen, crowned with faint lunar bows in the midst of the dashing spray. But it is in the darkest nights, when storms are blowing and the agitated waves are phosphorescent, that the most impressive displays are made. Then the long range of ice-bluffs, faintly illumined, is seen stretching through the gloom in weird, unearthly splendor, luminous foam dashing against it, and against every drifting berg; and amid all this wild, auroral splendor ever and anon some huge new-born berg dashes the living water into a yet brighter foam, and the streaming torrents pouring from its sides are worn as robes of light, while they roar in awful accord with the roaring winds, deep calling unto deep, glacier to glacier, from fiord to fiord.

To the lover of wildness Alaska offers a glorious field for either work or rest: landscape beauty in a thousand forms, things great and small, novel and familiar, as wild and pure as paradise. Wander where you may, wildness ever fresh and ever beautiful meets you in endless variety: ice-laden mountains, hundreds of miles of them peaked and pinnacled and crowded together like trees in groves, and so high and so divinely clad in clouds and air that they seem to belong more to heaven than to earth; inland plains grassy and flowery, dotted with groves and extending like seas all around to the rim of the sky; lakes and streams shining and singing, outspread in sheets of mazy embroidery in untraceable, measureless abundance, brightening every landscape, and keeping the ground fresh and fruitful forever; forests of ever-greens growing close together like leaves of grass, girdling a thousand islands and mountains in glorious array; mountains that are monuments of the work of ice, mountains mounuments of volcanic fires; gardens filled with the fairest flowers, giving their fragrance to every wandering wind; and far to the north thousands of miles of ocean ice, now wrapped in fog, now glowing in sunshine through nightless days, and again shining in wintry

The Century Magazine (August, 1897)

splendor beneath the beams of the aurora—sea, land, and sky one mass of white radiance like a star. Storms, too, are here as wild and sublime in size and scenery as the landscapes beneath them, displaying the glorious pomp of clouds on the march over mountain and plain, the flight of the snow when all the sky is in bloom, trailing rain-floods, and the booming plunge of avalanches and icebergs and rivers in their rocky glens; while multitudes of wild animals and wild people, clad in feathers and furs, fighting, loving, getting a living, make all the wildness wilder. All this, and unspeakably more, lies in wait for those who love it, sufficient in kind and quantity for gods and men. And notwithstanding that this vast wilderness with its wealth is in great part inaccessible to the streams of careworn people called "tourists," who go forth on ships and railroads to seek rest with nature once a year, some of the most interesting scenery in the territory has lately been brought within easy reach even of such travelers as these, especially in southeastern Alaska, where are to be fond the finest of the forests, the highest mountains, and the largest glaciers.

During the summer season good steamships carrying passengers leave Tacoma on Puget Sound for Alaska about once a week. After touching at Seattle, Port Townsend, Victoria, and Nanaimo, they go through a wilderness of islands to Wrangel, where the first stop in Alaska is made. Thence a charming, wavering course is pursued still north-

THE ALASKA TRIP

ward through the grandest scenery to Tahkou, Juneau, Chilcat, Glacier Bay, and Sitka, affording fine glimpses of the innumerable evergreen islands, the icy mountain-ranges of the coast, the forests, glaciers, etc. The round trip of two thousand miles is made in about twelve days, and costs about a hundred dollars: and though on ocean waters, there is no seasickness, for all the way lies through a network of sheltered inland channels and sounds that are about as free from heaving waves as rivers are.

No other excursion that I know of can be made into any of the wild portions of America where so much fine and grand and novel scenery is brought to view at so cheap and easy a price. Anybody may make this trip and be blest by it—old or young, sick or well, soft, succulent people whose limbs have never ripened, as well as sinewy mountaineers; for the climate is kindly, and one has only to breathe the exhilarating air and gaze and listen while being carried smoothly onward over the glassy waters. Even the blind may be benefited by laving and bathing in the balmy, velvety atmosphere, and the unjust as well as the just; for I fancy that even sins must be washed away in such a climate, and at the feet of such altars as the Alaska mountains are.

Between Tacoma and Port Townsend you gain a general view of the famous Puget Sound, for you sail down the middle of it. It is an arm and many-fingered hand of the sea reaching a hundred miles into the heart of one of the richest forest-regions on the globe. The scenery in fine weather is enchanting, the water as smooth and blue as a mountain lake, sweeping in beautiful curves around bays and capes and jutting promontories innumerable, and islands with soft, wavering outlines passing and overlapping one another, richly feathered with tall, spiry spruces, many of the trees 300 feet in height, their beauty doubled in reflections on the shiny waters. The Cascade Mountains bound the view on the right, the Olympic Range on the left, both ranges covered nearly to their summits with dense coniferous woods.

Doubling cape after cape, and passing uncounted islands that stud the shores, so many new and charming views are offered that one begins to feel there is no need of going farther. Sometimes clouds come down, blotting out all the land; then, lifting a little, perhaps a single island will be given back to the landscape, the tops of its trees dipping out of sight in trailing fringes of mist. Then the long ranks of spruce and cedar along the mainland are set free; and when at length the cloud-mantle vanishes, the colossal cone of Mount Rainier, 14,000 feet high, appears in spotless white, looking down over the dark woods like the very god of the landscape. A fine beginning is this for the Alaska trip! Crossing the Strait of Juan de Fuca from Port Townsend, in a few hours you are in Victoria and a foreign land. Victoria is a handsome little town, a section of old England set down nearly unchanged in the western American wilderness. It is situated on the south end of Vancouver Island, which is 280 miles long, the largest and southernmost of the wonderful archipelago that stretches northward along the margin of the continent for nearly a thousand miles. The steamer usually stops a few hours here, and most of the tourists go up town to the stores of the famous Hudson Bay Company to purchase fur or some wild Indian trinket as a memento. At certain seasons of the year, when the hairy harvests from the North have been gathered, immense bales of skins may be seen in the unsavory warehouses, the clothing of bears, wolves, beavers, otters, fishers, martens, lynxes, panthers, wolverenes, reindeer, moose, elk, wild sheep, foxes, seals, muskrats, and many others of "our poor earth-born companions and fellow-mortals."

The wilderness presses close up to the town, and it is wonderfully rich and luxuriant. The forests almost rival those of Puget Sound; wild roses are three inches in diameter, and ferns ten feet high. And strange to say, all this exuberant vegetation is growing on moraine material that has been scarcely moved or modified in any way by postglacial agents. Rounded masses of hard, resisting rocks rise everywhere along the shore and in the woods, their scored and polished surfaces still unwasted, telling of a time, so lately gone, when the whole region lay in darkness beneath an all-embracing mantle of ice. Even in the streets of the town glaciated bosses are exposed, the telling inscriptions of which have not been effaced by the wear of either weather or travel. And in the orchards fruitful boughs shade the edges of glacial pavements, and drop apples and peaches on them. Nowhere, as far as I have seen, are the beneficent influences of glaciers made manifest in plainer terms or with more striking contrasts. No tale of enchantment is so marvelous, so exciting to the imagination, as the story of the works and ways of snowflowers marching forth from their encampments on the mountains to develop the beauty of land scenes and make them fruitful.

Leaving Victoria, instead of going to sea we go into a shady wilderness that looks as though it might be in the heart of the continent. Most of the channels through which we glide are narrow as compared with their length and with the height of the mountain walls of the islands which bound their shores. But however sheer the walls, they are almost everywhere densely forested from the water's edge to a height of two thousand feet; and almost every tree may be seen as they rise above one another like an audience on a gallery—the blue-green, sharply spired Menzies spruce; the warm, yellow-green Merten spruce, with finger-like tops all pointing in the same direction or gracefully drooping; and the airy, feathery, brownish Alaska cedar. Most of the way we seem to be tracing a majestic river with lake-like expansions, the tide-currents, the fresh driftwood brought down by avalanches, the inflowing torrents, and the luxuriant foliage of the shores making the likeness complete. The steamer is often so near the shore that we can see the purple cones on the top branches of the trees, and the ferns and bushes at their feet. Then, rounding some bossy cape, the eye perchance is called away into a far-reaching vista, headlands on each side in charming array, one dipping gracefully beyond the other and growing finer in the distance, while the channel, like a strip of silver, stretches between, stirred here and there by leaping salmon and flocks of gulls and ducks that float like lilies among the sun-spangles. While we may be gazing into the depths of this leafy ocean lane, the ship, turning suddenly to right or left, enters an open space, a sound decorated with small islands, sprinkled or clustered in forms and compositions such as nature alone can invent. The smallest of the islands are mere dots, but how beautiful they are! The trees growing on them seem like handfuls that have been culled from the neighboring woods, nicely sorted and arranged, and then set in the water to keep them fresh, the fringing trees leafing out like flowers against the rim of a vase.

The variety we find, both as to the contours and collocation of the islands, whether great or small, is chiefly due to differences in the composition and physical structure of the rocks out of which they are made, and the unequal amount of glaciation to which they have been subjected. All the islands of the archipelago, as well as the headlands and promontories of the mainland, have a rounded, over-rubbed, sandpapered appearance, a finish free from angles, which is produced by the grinding of an oversweeping, ponderous flood of ice.

FORT WRANGEL

Seven hundred miles of this scenery, and we arrive at Fort Wrangel, on Wrangel Island, near the mouth of the Stickeen River. It is a quiet, rugged, dreamy place of no particular number of inhabitants—a few hundreds of whites and Indians, more or less, sleeping in a bog in the midst of the purest and most delightful scenery on the continent. Baron Wrangel established a trading-post here about a hundred years ago, and the fort, a quadrangular stockade, was built by the United States shortly after the purchase of the territory; but in a few years it was abandoned and sold to private parties. Indians, mostly of the Stickeen tribe, occupy the two long, draggled ends of the town along the shore; the whites, numbering about fifty, the middle portion. Stumps and logs roughen its two crooked streets, each of these picturesque obstructions mossy and tufted with grass and bushes on account of the dampness of the climate.

On the arrival of the steamer, most of the passengers make haste to go ashore to see the curious totem-poles in front of the massive timber houses of the Indians, and to buy curiosities, chiefly silver bracelets hammered from dollars

and half-dollars and tastefully engraved by Indian
workmen; blankets better than those of civilization,
woven from the wool of wild goats and sheep;
carved spoons from the horns of these animals;
Shaman rattles, miniature totem-poles, canoes,
paddles, stone hatchets, pipes, baskets, etc. The
traders in these curious wares are mostly women
and children, who gather on the front platforms of
the half-dozen stores, sitting in their blankets,
seemingly careless whether they sell anything or
not, every other face blackened hideously, a naked
circle about the eyes and on the tip of the nose,
where the smut has been weathered off. The larger
girls and the young women are brilliantly arrayed
in ribbons and calico, and shine among the blackened and blanketed old crones like scarlet tanagers
in a flock of blackbirds. Besides curiosities, most of
them have berries to sell, red, yellow and blue,
fresh and dewy, and looking wondrous clean as
compared with the people. These Indians are proud
and intelligent, nevertheless, and maintain an air
of self-respect which no amount of raggedness and
squalor can wholly subdue.

Many canoes may be seen along the shore, all
fashioned alike, with long, beak-like sterns and
prows, the largest carrying twenty or thirty persons. What the mustang is to the Mexican vaquero
the canoe is to the Indian of the Alaska coast. They
skim over the glassy, sheltered waters far and near
to fish and hunt and trade, or merely to visit their
neighbors. Yonder goes a whole family, grand-

parents and all, the prow of their canoe blithely decorated with handfuls of the purple epilobium. They are going to gather berries, as the baskets show. Nowhere else in my travels, north or south, have I seen so many berries. The woods and meadows and open spaces along the shores are full of them—huckleberries of many species, salmon-berries, raspberries, blackberries, currants, and gooseberries, with fragrant strawberries and service-berries on the drier grounds, and cran-berries in the bogs, sufficient for every worm, bird, and human being in the territory, and thousands of tons to spare. The Indians at certain seasons, roving in merry bands, gather large quantities, beat them into paste, and then press the paste into square cakes and dry them for winter use, to be eaten as a kind of bread with their oily salmon. Berries alone, with the lavish bloom that belongs to them, are enough to show how fine and rich this Northern wilderness must be.

ALASKA WEATHER

The climate of all that portion of the coast that is bathed by the Japan current, extending from the southern boundary of the territory northward and westward to the island of Atoo, a distance of nearly twenty-five hundred miles, is remarkably bland, and free from extremes of heat and cold throughout the year. It is rainy, however; but the rain is of good quality, gentle in its fall, filling the fountains of the streams, and keeping the whole land fresh and fruitful, while anything more delightful than the shining weather after the rain— the great, round sun-days of June, July, and August—can hardly be found elsewhere. An Alaska midsummer day is a day without night. In the extreme northern portion of the territory the sun does not set for weeks, and even as far south as Sitka and Fort Wrangel it sinks only a few degrees below the horizon, so that the rosy colors of the evening blend with those of the morning, leaving no gap of darkness between. Nevertheless, the full day opens slowly. At midnight, from the middle point between the gloaming and the dawn, a low arc of light is seen stealing along the horizon, with gradual increase of height and span and intensity of tone, accompanied usually by red clouds, which make a striking advertisement of the sun's progress long before he appears above the mountain-tops. For several hours after sunrise everything in the landscape seems dull and uncommunicative. The clouds fade, the islands and the mountains, with ruffs of mist about them, cast ill-defined shadows, and the whole firmament changes to pale pearl-gray with just a trace of purple in it. But toward noon there is a glorious awakening. The cool haziness of the air vanishes, and the richer sunbeams, pouring from on high, make all the bays and channels shine. Brightly now play the round-topped ripples about the edges of the islands, and over many a plume-shaped streak between them, where the water is stirred by some passing breeze.

WILDERNESS ESSAYS

On the mountains of the mainland, and in the high-walled fiords that fringe the coast, still finer is the work of the sunshine. The broad white bosoms of the glaciers glow like silver, and their crystal fronts, and the multitude of icebergs that linger about them, drifting, swirling, turning their myriad angles to the sun, are kindled into a perfect blaze of irised light. The warm air throbs and wavers, and makes itself felt as a life-giving, energizing ocean embracing all the earth. Filled with ozone, our pulses bound, and we are warmed and quickened into sympathy with everything, taken back into the heart of nature, whence we came. We feel the life and motion about us, and the universal beauty: the tides marching back and forth with weariless industry, laving the beautiful shores, and swaying the purple dulse of the broad meadows of the sea where the fishes are fed; the wild streams in rows white with waterfalls, ever in bloom and ever in song, spreading their branches over a thousand mountains; the vast forests feeding on the drenching sunbeams, every cell in a whirl of enjoyment; misty flocks of insects stirring all the air; the wild sheep and goats on the grassy ridges above the woods, bears in the berry-tangles, mink and beaver and otter far back on many a river and lake; Indians and adventurers pursuing their lonely ways; birds tending their young— everywhere, everywhere, beauty and life, and glad, rejoicing action.

Through the afternoon all the way down to the

west the air seems to thicken and become soft, without losing its fineness. The breeze dies away, and everything settles into a deep, conscious repose. Then comes the sunset with its purple and gold—not a narrow arch of color, but oftentimes filling more than half the sky. The horizontal clouds that usually bar the horizon are fired on the edges, and the spaces of clear sky between them are filled in with greenish yellow and amber; while the flocks of thin, overlapping cloudlets are mostly touched with crimson, like the outleaning sprays of a maple-grove in the beginning of Indian summer; and a little later a smooth, mellow purple flushes the sky to the zenith, and fills the air, fairly steeping and transfiguring the islands and mountains, and changing all the water to wine.

According to my own observations, in the year 1879 about one third of the summer weather at Wrangel was cloudy, one third rainy, and one third clear. Rain fell on eighteen days in June, eight in July, and twenty in September. But on some of these days only a light shower fell, scarce enough to count, and even the darkest and most bedraggled of them all had a dash of late or early color to cheer them, or some white illumination about the noon hours, while the lowest temperature was about 50°, and the highest 75°.

It is only in late autumn and winter that grand, roaring storms come down and solidly fill all the hours of day and night. Most of them are steady, all-day rains with high winds. Snow on the lowlands is not uncommon, but it never falls to a great depth, or lies long, and the temperature is seldom more than a few degrees below the freezing-point. On the mountains, however, and back in the interior, the winter months are intensely cold— so cold that mercury may at times be used for bullets by the hunters, instead of lead.

EXCURSIONS ABOUT WRANGEL

By stopping over a few weeks at Fort Wrangel, and making excursions into the adjacent region, many near and telling views may be had of the noble forests, glaciers, streams, lakes, wild gardens, Indian villages, etc.; and as the Alaska steamers call here about once a week, you can go on northward and complete your round trip when you like.

THE FORESTS

Going into the woods almost anywhere, you have first to force a way through an outer tangle of *Rubus,* huckleberry, dogwood, and elder-bushes, and a strange woody plant, about six feet high, with limber, rope-like stems beset with thorns, and a head of broad, translucent leaves like the crown of a palm. This is the *Echino panax horrida,* or devil's-club. It is used by the Indians for thrashing witches, and, I fear, deserves both of its bad names. Back in the shady deeps of the forest the walking is comparatively free, and you will be charmed with the majestic beauty and grandeur of the trees, as well as with the solemn stillness and the beauty of the elastic carpet of golden mosses flecked and barred with the sunbeams that sift through the leafy ceiling.

The bulk of the forests of southeastern Alaska is made up of three species of conifers—the Menzies and Merten spruces, and the yellow cedar. These trees cover nearly every rod of the thousand islands, and the coast and the slopes of the mountains of the mainland to a height of about 2000 feet above the sea.

The Menzies spruce, or Sitka pine (*Picea Sitchensis*), is the commonest species. In the heaviest portions of the forest it grows to a height of 175 feet or more, with a diameter of from three to six feet, and in habit and general appearance resembles the Douglas spruce, so abundant about Puget Sound. The timber is tough, close-grained, white, and looks like pine. A specimen that I examined back of Fort Wrangel was a little over six feet in diameter inside the bark four feet above the ground, and at the time it was felled was about 500 years old. Another specimen, four feet in diameter, was 385 years old; and a third, a little less than five feet thick, had attained the good old age of 764 years without showing any trace of decay. I saw a raft of this spruce that had been brought to Wrangel from one of the neighboring islands, three of the logs of which were one hundred feet in length, and nearly two feet in diameter at the small ends. Perhaps half of all the trees in southeastern Alaska are of this species. Menzies, whose name is associated with this grand tree, was a Scotch botanist who accompanied Vancouver in his voyage of discovery to this coast a hundred years ago.

The beautiful hemlock-spruce (*Tsuga Mertensiana*) is more slender than its companion, but nearly as tall, and the young trees are more graceful and picturesque in habit. Large numbers of this species used to be cut down by the Indians for the astringent bark, which they pounded into meal for bread to be eaten with oily fish.

The third species of this notable group, *Chamoecyparis Nutkaensis*, called yellow cedar or Alaska cedar, attains a height of 150 feet and a diameter of from three to five feet. The branches are pinnate, drooping, and form beautiful light-green sprays like those of *Libocedrus*, but the foliage is finer and the plumes are more delicate. The wood of this noble tree is the best the country affords, and one of the most valuable of the entire Pacific coast. It is pale yellow, close-grained, tough, durable, and takes a fine polish. The Indians make their paddles and totem-poles of it, and weave matting and coarse cloth from the inner bark. It is also the favorite fire-wood. A yellow-cedar fire is worth going a long way to see. The flames rush up in a multitude of quivering, jagged-edged lances, displaying admirable enthusiasm, while the burning surfaces of the wood snap and crackle and explode and throw off showers of coals with such noise that conversation at such firesides is well-nigh impossible.

The durability of this timber is forcibly illustrated by fallen trunks that are perfectly sound after lying in the damp woods for centuries. Soon after these trees fall they are overgrown with moss, in which seeds lodge and germinate and grow up into vigorous saplings, which stand in a row on the backs of their dead ancestors. Of this company of young trees perhaps three or four will grow to full stature, sending down straddling roots on each side, and establishing themselves in the soil; and after they have reached an age of two or three hundred years, the downtrodden trunk on which they are standing, when cut into, is found as fresh in the heart as when it fell.

The species is found as far south as Oregon, and is sparsely distributed along the coast and through the islands as far north as Chilcat (latitude 59°). The most noteworthy of the other trees found in the southern portion of these forests, but forming only a small portion of the whole, is the giant arbor-vitae (*Thuja gigantea*). It is distributed all the way up the coast from California to about latitude 56°. It is from this tree that the Indians make their best canoes, some of them being large enough to carry fifty or sixty men. Of pine I have seen only one species (*Pinus contorta*), a few specimens of which, about fifty feet high, may be found on the margins of lakes and bogs. In the interior beyond the mountains it forms extensive forests. So also does *Picea alba*, a slender, spiry tree which attains a height of one hundred feet or more. I saw this species growing bravely on frozen ground on the banks of streams that flow into Kotzebue Sound, forming there the margin of the arctic forest.

In the cool cañons and fiords, and along the banks of the glaciers, a species of silver fir and the beautiful Paton spruce abound. The only hard-wood trees I have found in Alaska are birch, alder, maple, and wild apple, one species of each. They grow mostly about the margins of the main forests and back in the mountain cañons. The lively yellow-green of the birch gives pleasing variety to the colors of the conifers, especially on slopes of river-cañons with a southern exposure. In general views all the coast forests look dark in the middle ground and blue in the distance, while the foreground shows a rich series of gray and brown and yellow trees. In great part these colors are due to lichens which hang in long tresses from the limbs, and to mosses which grow in broad, nest-like beds on the horizontal palmate branches of the Menzies and Merten spruces. Upon these moss-bed gardens high in the air ferns and grasses grow luxuriantly, and even seedling trees five or six feet in height, presenting the curious spectacle of old, venerable trees holding hundreds of their children in their arms.

Seward expected Alaska to become the ship-yard of the world, and so perhaps it may. In the meantime, as good or better timber for every use still abounds in California, Oregon, Washington, and

THE ALASKA TRIP

British Columbia; and let us hope that under better management the waste and destruction that have hitherto prevailed in our forests will cease, and the time be long before our Northern reserves need to be touched. In the hands of nature these Alaska tribes of conifers are increasing from century to century as the glaciers are withdrawn. May they be saved until wanted for worthy use—so worthy that we may imagine the trees themselves willing to come down the mountains to their fate!

THE RIVERS

The most interesting of the excursions that may be made from Fort Wrangel is the one up the Stickeen River. Perhaps twenty or thirty of the Alaska streams may be called rivers, but not one of them all, from the mighty Yukon, 2000 miles long, to the shortest of the mountain torrents pouring white from the glaciers, has been fully explored. From St. Elias the coast mountains extend in a broad, lofty chain beyond the southern boundary of the territory, gashed by stupendous cañons, each of which carries a stream deep enough and broad enough to be called a river, though comparatively short, as the highest sources of most of them lie in the icy solitudes of the range within forty or fifty miles of the coast. A few, however, of this foaming brotherhood—the Chilcat, Chilcoot, Tahkou, Stickeen, and perhaps others—come from beyond the range, heading with the Mackenzie and Yukon.

The tributary cañons of the main-trunk cañons of all these streams are still occupied by glaciers which descend in glorious ranks, their massy, bulging snouts lying back a little distance in the shadows, or pushed grandly forward among the cottonwoods that line the banks of the rivers, or all the way across the main cañons, compelling the rivers to find a way beneath them through long, arching tunnels.

The Stickeen is perhaps better known than any other river in Alaska, because it is the way to the Cassiar gold-mines. It is about 350 miles long, and is navigable for small steamers 150 miles to Glenora. It first pursues a westerly course through grassy plains darkened here and there with patches of evergreens; then, curving southward, and receiving numerous tributaries from the north, it enters the Coast Range, and sweeps across it to the sea, through a yosemite that is more than a hundred miles long, one to three miles wide and from 5000 to 8000 feet deep, and marvelously beautiful from end to end. To the appreciative tourist sailing up the river, the cañon is a gallery of sublime pictures, an unbroken series of majestic mountains, glaciers, waterfalls, cascades, groves, gardens, grassy meadows, etc., in endless variety of form and composition; while back of the walls, and thousands of feet above them, innumerable peaks and spires and domes of ice and snow tower grandly into the sky.

Gliding along the swift-flowing river, the views

change with bewildering rapidity. Wonderful, too, are the changes dependent on the seasons and the weather. In winter avalanches from the snow-laden heights boom and reverberate from side to side like majestic waterfalls; storm-winds from the arctic highlands, sweeping the cañon like a flood, choke the air with ice-dust; while the rocks, glaciers, and groves are in spotless white. In spring you enjoy the chanting of countless waterfalls; the gentle breathing of warm winds; the opening of leaves and flowers; the humming of bees over beds of honeybloom; birds building their nests; clouds of fragrance drifting hither and thither from miles of wild roses, clover, and honeysuckle, and tangles of sweet chaparral; swaths of birch and willow on the lower slopes following the melting snow-banks; bossy cumuli swelling in white and purple piles above the highest peaks; gray rain-clouds wreath-ing the outstanding brows and battlements of the walls; then the breaking forth of the sun after the rain, the shining of the wet leaves and the river and the crystal architecture of the glaciers; the rising of fresh fragrance, the song of the happy birds, the looming of the white domes in the azure, and the serene color-grandeur of the morning and evening. In summer you find the groves and gardens in full dress; glaciers melting rapidly under warm sunshine and rain; waterfalls in all their glory; the river rejoicing in its strength; butterflies wavering and drifting about like ripe flower-bloom in springtime; young birds trying their wings;

bears enjoying salmon and berries; all the life of the cañon brimming full like the streams. In autumn comes rest, as if the year's work were done; sunshine, streaming over the cliffs in rich, hazy beams, calls forth the last of the gentians and goldenrods; the groves and tangles and meadows bloom again, every leaf changing to a petal, scarlet and yellow; the rocks also bloom, and the glaciers, in the mellow golden light. And so goes the song, change succeeding change in glorious harmony through all the seasons and years.

Leaving Wrangel, you go up the coast to Juneau. After passing through the picturesque Wrangel Narrows into Souchoi Channel and Prince Frederick Sound, a few icebergs come in sight, the first you have seen on the trip. They are derived from a large, showy glacier, the Leconte, which discharges into a wild fiord near the mouth of the Stickeen River, which the Indians call *Hutli*, or Thunder Bay, on account of the noise made by the discharge of the icebergs. This, so far as I know, is the southernmost of the glaciers that flow into the sea. Gliding northward, you have the mountains of the mainland on one hand, Kuprianof and countless smaller islands on the other. The views extend far into the wilderness, all of them as wild and clean as the sky; but your attention will chiefly be turned to the mountains, now for the first time appreciably near. As the steamer crawls along the coast, the cañons are opened to view and closed again in regular succession, like the leaves of a book, allowing the attentive observer to see far back into their icy depths. About halfway between Wrangel Narrows and Cape Fanshaw, you are opposite a noble group of glaciers which come sweeping down through the woods from their white fountains nearly to the level of the sea, swaying in graceful, river-like curves around the feet of lofty granite mountains and precipices like those of the Yosemite valley. It was at the largest of these, the Paterson glacier, that the ships of the Alaska Ice Company were loaded for San Francisco and the Sandwich Islands.

An hour or two farther north another fleet of icebergs come in sight, which have their sources in Sum Dum or Holkam Bay. This magnificent inlet, with its long, icy arms reaching deep into the mountains, is one of the most interesting of all the Alaska fiords; but the icebergs in it are too closely compacted to allow a passage for any of the excursion-steamers.

About five miles from the mouth the bay divides into two main arms, about eighteen and twenty miles long, in the farthest-hidden recesses of which there are four large glaciers which discharge bergs. Of the smaller glaciers of the second and third class that melt before reaching tidewater, a hundred or more may be seen along the walls from a canoe, and about as many snowy cataracts, which, with the plunging bergs from the main glaciers, keep all the fiord in a roar. The scenery in both of the long arms and their side branches is of the wildest description, especially in their upper reaches, where the granite walls rise in sheer, massive precipices, like those of the Yosemite valley, to a height of from 3000 to 5000 feet. About forty miles farther up the coast another fleet of icebergs come in sight, through the midst of which the steamer passes into the Tahkou Inlet. It is about eighteen miles into the heart of the Coast Mountains, draining many glaciers, great and small, all of which were once tributary branches of one grand glacier that formed and occupied the inlet as its channel. This inlet more plainly than any other that I have examined illustrates the mode of formation of the wonderful system of deep channels extending northward from Puget Sound; for it is a marked portion of that system, a branch of Stephen's Passage still in process of formation at the head; while its trends and sculpture are as distinctly glacial as those of the smaller fiords.

Sailing up the middle of it, you may count some forty-five glaciers. Three of these reach the level of the sea, descending from a group of lofty mountains at the head of the inlet, and making a grand show. Only one, however, the beautiful Tahkou glacier, discharges bergs. It comes sweeping forward in majestic curves, and discharges its bergs through a western branch of the inlet next the one occupied by the Tahkou River. Thus we see here a river of ice and a river of water flowing into the sea side by side, both of them abounding in cascades and rapids; yet how different in their rate of motion, and in the songs they sing, and in their influence on the landscape! A rare object-lesson this, worth coming round the world to see.

Once, while I sat sketching among the icebergs here, two Tahkou Indians, father and son, came gliding toward us in an exceedingly small cottonwood canoe. Coming alongside with a good-natured "Sahgaya," they inquired who we were, what we were doing, etc., while they in turn gave information concerning the river, their village, and two other large glaciers a few miles up the river-cañon. They were hunting hair-seals, and as they slipped softly away in pursuit of their prey, crouching in their tiny shell of a boat among the bergs, with barbed spear in place, they formed a picture of icy wildness as telling as any to be found amid the drifts and floes of Greenland.

After allowing the passengers a little time— half an hour or so—to admire the crystal wall of the great glacier and the huge bergs that plunge and rise from it, the steamer goes down the inlet to Juneau. This young town is the mining-center, and, so far as business is concerned, the chief place in the territory. Here, it is claimed, you may see the largest quartz-mill in the world, the two hundred and forty stamps of which keep up a "steady, industrious growl that may be heard a mile away."

Alaska, generally speaking, is a hard country for the prospector, because most of the ground is either permanently frozen or covered with glaciers, forests, or a thick blanket of moss. Nevertheless, thousands of hardy miners from the gulches and ledges of California and Arizona are rapidly over-running the territory in every direction, and making it tell its wealth. And though perhaps not one vein or placer in a hundred has yet been touched, enough has been discovered to warrant the opinion that this icy country holds at least a fair share of the gold of the world. After time has been given for a visit to the mines and a saunter through the streets of Juneau, the steamer passes between Douglass and Admiralty islands into Lynn Canal, the most beautiful and spacious of all the mountain-walled channels you have yet seen. The Auk and Eagle glaciers appear in one view on the right as you enter the canal, swaying their crystal floods through the woods with grand effect. But it is on the west side of the canal, near the head, that the most striking feature of the landscape is seen—the Davidson glacier. It first appears as an immense ridge of ice thrust forward into the channel; but when you have gained a position directly in front, it presents a broad current issuing from a noble gateway at the foot of the mountains, and spreading out to right and left in a beautiful fanshaped mass three or four miles in width, the front of which is separated from the water by the terminal moraine. This is one of the most notable of the large glaciers that are in the first stage of decadence, reaching nearly to tide-water, but failing to enter it and send off bergs. Excepting the Tahkou, all the great glaciers you have yet seen on

THE ALASKA TRIP

the trip belong to this class; but this one is perhaps the most beautiful of its kind, and you will not be likely to forget the picture it makes, however icy your after-travels may be. Shortly after passing the Davidson glacier the northernmost point of the trip is reached at the head of the canal, a little above latitude 59°. At the canning-establishments here you may learn something of the inhabitants of these beautiful waters. Whatever may be said of other resources of the territory,—furs, minerals, timber, etc.,—it is hardly possible to overestimate the importance of the fisheries. Besides whales in the far North, and the cod, herring, halibut, and other food fishes that swarm over immense areas along the shores and inlets, there are probably not fewer than a thousand salmon streams in Alaska that are crowded with fine salmon for months every year. Their numbers are beyond conception. Oftentimes there seem to be more fish than water in the rapid portions of the streams. On one occasion one of my men waded out into the middle of a crowded run, and amused himself by picking up the fish and throwing them over his head. In a single hour these Indians may capture enough to last a year. Surely in no part of the world may one's daily bread be more easily obtained. Sailing into these streams on dark nights, when the waters are phosphorescent and the salmon are running, is a very beautiful and exciting experience; the myriad fins of the on-rushing multitude crowding against one another churn all the water from bank to bank into silver fire, making a glorious glow in the darkness.

From Chilcat we now go down Lynn Canal, through Icy Strait, and into the famous Glacier Bay. All the voyage thus far after leaving Wrangel has been icy, and you have seen hundreds of glaciers great and small; but this bay, and the region about it and beyond it toward Mount St. Elias, are pre-eminently the iceland of Alaska, and of all the west coast of the continent.

GLACIERS OF THE PACIFIC COAST

Glancing for a moment at the results of a general exploration of the mountain-ranges of the Pacific coast, we find that there are between sixty and seventy small residual glaciers in the California Sierra. Northward through Oregon and Washington, glaciers, some of them of considerable extent, still exist on all the higher volcanic mountains of the Cascade Range,—the Three Sisters, Mounts Jefferson, Hood, St. Helen's, Adams, Rainier, Baker, and others,—though none of them approach the sea. Through British Columbia and southeastern Alaska the broad, sustained chain of coast mountains is generally glacier-bearing. The upper branches of nearly every one of its cañons are still occupied by glaciers, which gradually increase in size and descend lower until the lofty region between Glacier Bay and Mount St. Elias is reached, where a considerable number discharge into the sea. About Prince William's Sound and Cook's Inlet many grand glaciers are displayed; but farther to the west, along the Alaska peninsula and the chain of the Aleutian Islands, though a large number of glaciers occur on the highest peaks, they are mostly small, and melt far above sea-level, while to the north of latitude 62° few, if any, remain in existence, the ground being comparatively low and the snowfall light.

ON THE MUIR GLACIER

The largest of the seven glaciers that discharge into Glacier Bay is the Muir; and being also the most accessible, it is the one to which tourists are taken and allowed to go ashore for a few hours, to climb about its crystal cliffs and watch the huge icebergs as with tremendous, thundering roar they plunge and rise from the majestic frontal sea-wall in which the glacier terminates. The front, or snout, of the glacier is about three miles wide, but the central berg-discharging portion, which stretches across from side to side of the inlet like a huge jagged white-and-blue barrier, is only about half as wide. The height of the ice-wall above the water is from 250 to 300 feet, but soundings made by Captain Carroll show that 720 feet of the wall is below the surface, while still a third unmeasured portion is buried beneath the moraine material that is being constantly deposited at the foot of it. Therefore, were the water and rocky detritus removed, there would be presented a sheer precipice of ice a mile and a half wide and more than a thousand feet in height. Seen from the inlet as you approach it, at a distance of a mile or two it seems massive and comparatively regular in form, but it is far from being smooth. Deep rifts and hollows alternate with broad, plain bastions, which are ever changing as the icebergs are discharged, while it is roughened along the top with innumerable spires and pyramids and sharp, hacked blades, leaning and toppling, or cutting straight into the sky.

THE BIRTH OF THE ICEBERGS

The number of bergs given off varies somewhat with the weather and the tides. For twelve consecutive hours I counted the number discharged that were large enough to make themselves heard like thunder at a distance of a mile or two, and found the average rate to be one in five or six minutes. The thunder of the largest may be heard, under favorable circumstances, ten miles or more. When a large mass sinks from the upper fissured portion of the wall, there is first a keen, piercing crash, then a deep, deliberate, long-drawn-out, thundering roar, which slowly subsides into a comparatively low, far-reaching, muttering growl; then come a crowd of grating, clashing sounds from the agitated bergs that dance in the waves about the newcomer as if in welcome; and these, again, are followed by the swash and roar of the berg-waves as they reach the shore and break among the boulders. But the largest and most beautiful of the bergs, instead of falling from the exposed weathered portion of the wall, rise from the submerged portion with a still grander commotion, heaving aloft nearly to the top of the wall with awful roaring, tons of water streaming like hair down their sides, while they heave and plunge again and again before they settle in poise and sail away as blue crystal islands, free at last after being held fast as part of a slow-crawling glacier for centuries. And how wonderful it seems that ice formed from pressed snow on the mountains two or three hundred years ago should, after all its toil and travel in grinding down and fashioning the face of the landscape, still remain pure and fresh and lovely in color! When the sunshine is pouring and sifting in iris colors through the midst of all this wilderness of angular crystal ice, and through the grand, flame-shaped jets and sheets of radiant spray ever rising from the blows of the falling bergs, the effect is indescribably glorious.

GLACIAL NIGHTS

Glorious, too, are the nights along these crystal cliffs, when the moon and the stars are shining; the projecting buttresses and battlements, seemingly far higher than by day, standing forward in the moonlight, relieved by the shadows of the hollows; the new-born bergs keeping up a perpetual storm of thunder, and the lunar bows displaying faint iris colors in the up-dashing spray. But it is in the darkest nights, when storms are blowing and the waters of the inlet are phosphorescent, that the most terribly impressive show is displayed. Then the long range of crystal bluffs, faintly illumined, is seen stretching away in the stormy gloom in awful, unearthly grandeur, luminous waves dashing beneath in a glowing, seething, wavering fringe of foam, while the new-born bergs, rejoicing in their freedom, plunging, heaving, grating one against another, seem like living creatures of some other world, dancing and roaring with the roaring storm and the glorious surges of auroral light.

CHARACTERISTICS OF THE MUIR GLACIER

If you go ashore as soon as the steamer drops anchor, you will have time to push back across the terminal moraine on the east side, and over a mile or so of the margin of the glacier, climb a yellow ridge that comes forward there and is easy of access, and gain a good, comprehensive, telling view of the greater portion of the glacier and its principal tributaries—that is, if you are so fortunate as to have clear weather. Instead of a river of ice winding down a narrow, mountain-walled valley, like the largest of the Swiss glaciers, you will see here a grand lake or sea of ice twenty-five or thirty miles wide, more than two hundred times as large as the celebrated Mer de Glace of the Alps, a broad, gently undulating prairie surrounded by a forest of mountains from the shadowy cañons and amphitheaters of which uncounted tributary glaciers flow into the grand central reservoir. There are

THE ALASKA TRIP

seven main tributaries, from two to six miles wide
where they enter the trunk, and from twenty to
thirty miles long; each of these has many secondary tributaries, so that the whole number, great
and small, pouring from the mountain fountains
into the grand central trunk must number at least
two hundred, not counting the smallest. The views
up the main tributaries in bright weather are exceedingly rich and beautiful; though far off from
your standpoint, the broad white floods of ice are
clearly seen issuing in graceful lines from the
depths of the mysterious solitudes. The area
drained by this one grand glacier and its branches
can hardly be less than a thousand square miles,
and it probably contains more ice than all the
eleven hundred glaciers of the Swiss Alps combined. The distance back from the front to the
head of the farthest fountain is about fifty miles,
and the width of the trunk below the confluence
of the tributaries is about twenty-five miles.
Though apparently as motionless as the mountains
about its basin, the whole glacier flows on like a
river, unhalting, unresting, through all the seasons
from century to century, with a motion varying in
every part with the depth of the current and the
declivity, smoothness, and directness of different
portions of the channel. The rate of motion in the
central cascading portion of the current near the
front, as determined by Professor Reid, is from
two and a half to five inches an hour, or from five
to ten feet a day.

WILDERNESS ESSAYS

Along the eastern margin of the main trunk the ice is so little broken that a hundred horsemen might ride abreast for miles without encountering much difficulty. But far the greater portion of the vast expanse is torn and crumpled into a bewildering network of ridges and blades, and rough, broken hummocks, separated by yawning gulfs and crevasses unspeakably beautiful and awful. Here and there the adventurous explorer, picking a way in long, patient zigzags through the shining wilderness, comes to spacious hollows, some of them miles in extent, where the ice, closely pressed and welded, presents beautiful blue lakes fed by bands of streams that sing and ring and gurgle, and make sheets of melody as sweet as ever were made by larks in springtime over their nests in the meadows.

Besides the Muir there are here six other noble glaciers which send off fleets of icebergs, and keep the whole bay in a roar. These are the Geikie, Hugh Miller, Pacific, Reid, Carroll, and Hoona glaciers. Of the second class of grand size descending to the level of the sea, but separated from it by mud floats and flood-washed terminal moraines, there are eight, and the smaller ones are innumerable.

With these views of the ice-world the duty-laden tourist is gladly content, knowing that nowhere else could he have sailed in a comfortable steamer into new-born landscapes and witnessed the birth of icebergs. Returning down the bay in a zigzag course, dodging the drifting bergs, you may

THE ALASKA TRIP

see the lofty summits of the Fairweather Range—Mounts Fairweather, Lituya, Crillon, and La Pérouse. Then, leaving Icy Strait, you enter Chatham Strait, and thence pass through the picturesque Peril Strait to Sitka, the capital of the territory. Here the steamer usually stops for a day, giving time to see the interesting old Russian town and its grand surroundings. After leaving Sitka the steamer touches again at Wrangel for the mails. Then, gliding through the green archipelago by the same way that you came you speedily arrive in civilization, rich in wildness forevermore.

TWENTY HILL HOLLOW

I WISH to say a word for the great central plain of California in general, and for Twenty Hill Hollow, in Merced County, in particular; because, in reading descriptions of California scenery, by the literary racers who annually make a trial of their speed here, one is led to fancy, that, outside the touristical see-saw of Yosemite, Geysers, and Big Trees, our State contains little else worthy of note, excepting, perhaps, certain wine-cellars and vineyards, and that our great plain is a sort of Sahara, whose narrowest and least dusty crossing they benevolently light-house. But to the few travelers who are in earnest — true lovers of the truth and beauty of wildness — we would say, Heed nothing you have heard; put no questions to "agent," or guide-book, or dearest friend; cast away your watches and almanacs, and go at once to our garden-

Overland Monthly (April, 1872)

This is the hub of the region where Mr. Muir spent the greater part of the summer of 1868 and the spring of 1869.

wilds — the more planless and ignorant the better. Drift away confidingly into the broad gulf-streams of Nature, helmed only by Instinct. No harsh storm, no bear, no snake, will harm you. Those who submissively allow themselves to be packed and brined down in the sweats of a stage-coach, who are hurled into Yosemite by "favorite routes," are not aware that they are crossing a grander Yosemite than that to which they are going.

The whole State of California, from Siskiyou to San Diego, is one block of beauty, one matchless valley; and our great plain, with its mountain-walls, is the true California Yosemite — exactly corresponding in its physical character and proportions to that of the Merced. Moreover, as Yosemite the less is outlined in the lesser Yosemites of Indian Cañon, Glacier Cañon, Illilouette, and Pohono, so is Yosemite the great by the Yosemites of King's River, Fresno, Merced, and Tuolumne. The only important difference between the great central Yosemite — bottomed by the plain of the Sacramento and San Joaquin, and walled by the Sierras and mountains of the coast — and the Merced Yosemite — bottomed by a glacier meadow, and walled by glacier rocks — is, that the former is double — two Yosemites in one, each proceeding from a tangle of glacier *cañons*, meeting opposite Suisun Bay, and sending their united waters to the sea by the Golden Gate.

Were we to cross-cut the Sierra Nevada into blocks a dozen miles or so in thickness, each section would contain a Yosemite Valley and a river, together with a bright array of lakes and meadows, rocks and forests. The grandeur and inexhaustible beauty of each block would be so vast and over-satisfying that to choose among them would be like selecting slices of bread cut from the same loaf. One bread-slice might have burnt spots, answering to craters; another would be more browned; another, more crusted or raggedly cut; but all essentially the same. In no greater degree would the Sierra slices differ in general character. Nevertheless, we all would choose the Merced slice, because, being easier of access, it has been nibbled and tasted, and pronounced very good; and because of the concentrated form of its Yosemite, caused by certain conditions of baking, yeasting, and glacier-frosting of this portion of the great Sierra loaf. In like manner, we readily perceive that the great central plain is one batch of bread — one golden cake — and we are loath to leave these magnificent loaves for crumbs, however good.

After our smoky sky has been washed in the rains of winter, the whole complex row of Sierras appear from the plain as a simple wall slightly beveled, and colored in horizontal bands laid one above another, as if entirely composed of partially straightened rainbows. So, also, the plain seen from the mountains has the same simplicity of smooth surface, colored purple and yellow, like a patchwork of irised clouds. But when we descend to this smooth-furred sheet, we discover complexity in its physical conditions equal to that of the mountains, though less strongly marked. In particular, that portion of the plain lying between the Merced and the Tuolumne, within ten miles of the slaty foothills, is most elaborately carved into valleys, hollows, and smooth undulations, and among them is laid the Merced Yosemite of the plain — Twenty Hill Hollow.

This delightful Hollow is less than a mile in length, and of just sufficient width to form a well-proportioned oval. It is situated about midway between the two rivers, and five miles from the Sierra foothills. Its banks are formed of twenty hemispherical hills; hence its name. They surround and enclose it on all sides, leaving only one narrow opening

toward the southwest for the escape of its
waters. The bottom of the Hollow is about
two hundred feet below the level of the surrounding plain, and the tops of its hills are
slightly below the general level. Here is no
towering dome, no Tissiack, to mark its
place; and one may ramble close upon its rim
before he is made aware of its existence. Its
twenty hills are as wonderfully regular in
size and position as in form. They are like
big marbles half buried in the ground, each
poised and settled daintily into its place at
a regular distance from its fellows, making
a charming fairy-land of hills, with small,
grassy valleys between, each valley having a
tiny stream of its own, which leaps and sparkles
out into the open hollow, uniting to form
Hollow Creek.

Like all others in the immediate neighborhood, these twenty hills are composed of
stratified lavas mixed with mountain drift
in varying proportions. Some strata are almost
wholly made up of volcanic matter — lava and
cinders — thoroughly ground and mixed
by the waters that deposited them; others
are largely composed of slate and quartz
boulders of all degrees of coarseness, forming conglomerates. A few clear, open sections
occur, exposing an elaborate history of seas,

and glaciers, and volcanic floods — chapters of cinders and ashes that picture dark days, when these bright snowy mountains were clouded in smoke, and rivered and laked with living fire. A fearful age, say mortals, when these bright Sierras flowed lava to the sea. What horizons of flame! What atmospheres of ashes and smoke!

The conglomerates and lavas of this region are readily denuded by water. In the time when their parent sea was removed to form this golden plain, their regular surface, in great part covered with shallow lakes, showed little variation from motionless level until torrents of rain and floods from the mountains gradually sculptured the simple page to the present diversity of bank and brae, creating, in the section between the Merced and the Tuolumne, Twenty Hill Hollow, Lily Hollow, and the lovely valley of Cascade and Castle Creeks, with many others nameless and unknown, seen only by hunters and shepherds, sunk in the wide bosom of the plain, like undiscovered gold. Twenty Hill Hollow is a fine illustration of a valley created by erosion of water. Here are no Washington columns, no angular El Capitans. The hollow cañons, cut in soft lavas, are not so deep as to require a single earthquake at the hands of science, much less a baker's dozen of those convenient tools demanded for the making of mountain Yosemites, and our moderate arithmetical standards are not outraged by a single magnitude of this simple, comprehensible hollow.

The present rate of denudation of this portion of the plain seems to be about one tenth of an inch per year. This approximation is based upon observations made upon stream-banks and perennial plants. Rains and winds remove mountains without disturbing their plant or animal inhabitants. Hovering petrels, the fishes and floating plants of ocean, sink and rise in beautiful rhythm with its waves; and, in like manner, the birds and plants of the plain sink and rise with these waves of land, the only difference being that the fluctuations are more rapid in the one case than in the other.

In March and April the bottom of the Hollow and every one of its hills are smoothly covered and plushed with yellow and purple flowers, the yellow predominating. They are mostly social *Compositæ*, with a few claytonias, gilias, eschscholtzias, white and yellow violets, blue and yellow lilies, dodecatheons, and eriogonums set in a half-floating maze of purple grasses. There is but one vine in the Hollow — the *Megarrhiza* [*Echinocystis* T. & D.] or

TWENTY HILL HOLLOW

"Big Root." The only bush within a mile of it, about four feet in height, forms so remarkable an object upon the universal smoothness that my dog barks furiously around it, at a cautious distance, as if it were a bear. Some of the hills have rock ribs that are brightly colored with red and yellow lichens, and in moist nooks there are luxuriant mosses — *Bartramia*, *Dicranum*, *Funaria*, and several *Hypnums*. In cool, sunless coves the mosses are companioned with ferns — a *Cystopteris* and the little gold-dusted rock fern, *Gymnogramma triangularis*.

The Hollow is not rich in birds. The meadow-lark homes there, and the little burrowing owl, the killdeer, and a species of sparrow. Occasionally a few ducks pay a visit to its waters, and a few tall herons — the blue and the white — may at times be seen stalking along the creek; and the sparrow hawk and gray eagle [1] come to hunt. The lark, who does nearly all the singing for the Hollow, is not identical in species with the meadowlark of the East, though closely resembling it; richer flowers and skies have inspired him with a better song than was ever known to the Atlantic lark.

I have noted three distinct lark-songs here. The words of the first, which I committed to

[1] Mr. Muir doubtless meant the golden eagle (*Aquila chrysaëtos*).

memory at one of their special meetings, spelled as sung, are "Wee-ro spee-ro wee-o weer-ly wee-it." On the 20th of January, 1869, they sang "Queed-lix boodle," repeating it with great regularity, for hours together, to music sweet as the sky that gave it. On the 22d of the same month, they sang "Chee chool chee-dildy choodildy." An inspiration is this song of the blessed lark, and universally absorbable by human souls. It seems to be the only bird-song of these hills that has been created with any direct reference to us. Music is one of the attributes of matter, into whatever forms it may be organized. Drops and sprays of air are specialized, and made to plash and churn in the bosom of a lark, as infinitesimal portions of air plash and sing about the angles and hollows of sand-grains, as perfectly composed and pre-destined as the rejoicing anthems of worlds; but our senses are not fine enough to catch the tones. Fancy the waving, pulsing melody of the vast flower-congregations of the Hollow flowing from myriad voices of tuned petal and pistil, and heaps of sculptured pollen. Scarce one note is for us; nevertheless, God be thanked for this blessed instrument hid beneath the feathers of a lark.

The eagle does not dwell in the Hollow; he only floats there to hunt the long-eared hare.

TWENTY HILL HOLLOW

One day I saw a fine specimen alight upon a hillside. I was at first puzzled to know what power could fetch the sky-king down into the grass with the larks. Watching him attentively, I soon discovered the cause of his earthiness. He was hungry and stood watching a long-eared hare, which stood erect at the door of his burrow, staring his winged fellow mortal full in the face. They were about ten feet apart. Should the eagle attempt to snatch the hare, he would instantly disappear in the ground. Should long-ears, tired of inaction, venture to skim the hill to some neighboring burrow, the eagle would swoop above him and strike him dead with a blow of his pinions, bear him to some favorite rock table, satisfy his hunger, wipe off all marks of grossness, and go again to the sky.

Since antelopes have been driven away, the hare is the swiftest animal of the Hollow. When chased by a dog he will not seek a burrow, as when the eagle wings in sight, but skims wavily from hill to hill across connecting curves, swift and effortless as a bird-shadow. One that I measured was twelve inches in height at the shoulders. His body was eighteen inches, from nose-tip to tail. His great ears measured six and a half inches in length and two in width. His ears — which, notwithstanding their great size, he wears gracefully and becomingly — have procured for him the homely nickname, by which he is commonly known, of "Jackass rabbit." Hares are very abundant over all the plain and up in the sunny, lightly wooded foothills, but their range does not extend into the close pine forests.

Coyotes, or California wolves, are occasionally seen gliding about the Hollow; but they are not numerous, vast numbers having been slain by the traps and poisons of sheep-raisers. The coyote is about the size of a small shepherd-dog, beautiful and graceful in motion, with erect ears, and a bushy tail, like a fox. Inasmuch as he is fond of mutton, he is cordially detested by "sheep-men" and nearly all cultured people.

The ground-squirrel is the most common animal of the Hollow. In several hills there is a soft stratum in which they have tunneled their homes. It is interesting to observe these rodent towns in time of alarm. Their one circular street resounds with sharp, lancing outcries of "Seekit, seek, seek, seekit!" Near neighbors, peeping cautiously half out-of-doors, engage in low, purring chat. Others, bolt upright on the doorsill or on the rock above, shout excitedly, as if calling attention to the motions and aspects of the enemy. Like the wolf, this little animal is accursed, because of his relish for grain. What a pity that Nature should have made so many small mouths palated like our own!

All the seasons of the Hollow are warm and bright, and flowers bloom through the whole year. But the grand commencement of the annual genesis of plant and insect life is governed by the setting-in of the rains, in December or January. The air, hot and opaque, is then washed and cooled. Plant seeds, which for six months have lain on the ground dry as if garnered in a farmer's bin, at once unfold their treasured life. Flies hum their delicate tunes. Butterflies come from their coffins, like cotyledons from their husks. The network of dry water-courses, spread over valleys and hollows, suddenly gushes with bright waters, sparkling and pouring from pool to pool, like dusty mummies risen from the dead and set living and laughing with color and blood. The weather grows in beauty, like a flower. Its roots in the ground develop day-clusters a week or two in size, divided by and shaded in foliage of clouds; or round hours of ripe sunshine wave and spray in sky-shadows, like racemes of berries half hidden in leaves.

These months of so-called rainy season are not filled with rain. Nowhere else in North

America, perhaps in the world, are Januarys so balmed and glowed with vital sunlight. Referring to my notes of 1868 and 1869, I find that the first heavy general rain of the season fell on the 18th of December. January yielded to the Hollow, during the day, only twenty hours of rain, which was divided among six rainy days. February had only three days on which rain fell, amounting to eighteen and one half hours in all. March had five rainy days. April had three, yielding seven hours of rain. May also had three wet days, yielding nine hours of rain, and completed the so-called "rainy season" for that year, which is probably about an average one. It must be remembered that this rain record has nothing to do with what fell in the night.

The ordinary rainstorm of this region has little of that outward pomp and sublimity of structure so characteristic of the storms of the Mississippi Valley. Nevertheless, we have experienced rainstorms out on these treeless plains, in nights of solid darkness, as impressively sublime as the noblest storms of the mountains. The wind, which in settled weather blows from the northwest, veers to the southeast; the sky curdles gradually and evenly to a grainless, seamless, homogeneous cloud; and then comes the rain, pouring steadily and often driven aslant by strong winds. In 1869, more than three fourths of the winter rains came from the southeast. One magnificent storm from the northwest occurred on the 21st of March; an immense, round-browed cloud came sailing over the flowery hills in most imposing majesty, bestowing water as from a sea. The passionate rain-gush lasted only about one minute, but was nevertheless the most magnificent cataract of the sky mountains that I ever beheld. A portion of calm sky toward the Sierras was brushed with thin, white cloud-tissue, upon which the rain-torrent showed to a great height — a cloud waterfall, which, like those of Yosemite, was neither spray, rain, nor solid water. In the same year the cloudiness of January, omitting rainy days, averaged 0.32; February, 0.13; March, 0.20; April, 0.10; May, 0.08. The greater portion of this cloudiness was gathered into a few days, leaving the others blocks of solid, universal sunshine in every chink and pore.

At the end of January, four plants were in flower: a small white cress, growing in large patches; a low-set, umbelled plant, with yellow flowers; an eriogonum, with flowers in leafless spangles; and a small boragewort. Five or six mosses had adjusted their hoods, and were in the prime of life. In February, squirrels, hares, and flowers were in springtime joy. Bright plant-constellations shone everywhere about the Hollow. Ants were getting ready for work, rubbing and sunning their limbs upon the husk-piles around their doors; fat, pollen-dusted, "burly, dozing humble-bees" were rumbling among the flowers; and spiders were busy mending up old webs, or weaving new ones. Flowers were born every day, and came gushing from the ground like gayly dressed children from a church. The bright air became daily more songful with fly-wings, and sweeter with breath of plants.

In March, plant-life is more than doubled. The little pioneer cress, by this time, goes to seed, wearing daintily embroidered silicles. Several claytonias appear; also, a large white leptosiphon [?], and two nemophilas. A small plantago becomes tall enough to wave and show silky ripples of shade. Toward the end of this month or the beginning of April, plant-life is at its greatest height. Few have any just conception of its amazing richness. Count the flowers of any portion of these twenty hills, or of the bottom of the Hollow, among the streams: you will find that there are from one to ten thousand upon every square yard, counting the heads of *Compositæ* as single flowers. Yellow *Compositæ* form by far the greater portion

of this goldy-way. Well may the sun feed them
with his richest light, for these shining sunlets
are his very children — rays of his ray, beams
of his beam! One would fancy that these California days receive more gold from the ground
than they give to it. The earth has indeed become a sky; and the two cloudless skies, raying
toward each other flower-beams and sunbeams,
are fused and congolded into one glowing
heaven. By the end of April most of the Hollow plants have ripened their seeds and died;
but, undecayed, still assist the landscape with
color from persistent involucres and corolla-
like heads of chaffy scales.

In May, only a few deep-set lilies and eriogonums are left alive. June, July, August, and
September are the season of plant rest, followed, in October, by a most extraordinary out-
gush of plant-life, at the very driest time of the
whole year. A small, unobtrusive plant, *Hemi-
zonia virgata*, from six inches to three feet in
height, with pale, glandular leaves, suddenly
bursts into bloom, in patches miles in extent,
like a resurrection of the gold of April. I have
counted upward of three thousand heads upon
one plant. Both leaves and pedicels are so
small as to be nearly invisible among so vast
a number of daisy golden-heads that seem to
keep their places unsupported, like stars in the

sky. The heads are about five eighths of an inch in diameter; rays and disk-flowers, yellow; stamens, purple. The rays have a rich, furred appearance, like the petals of garden pansies. The prevailing summer wind makes all the heads turn to the southeast. The waxy secretion of its leaves and involucres has suggested its grim name of "tarweed," by which it is generally known. In our estimation, it is the most delightful member of the whole Composite Family of the plain. It remains in flower until November, uniting with an eriogonum that continues the floral chain across December to the spring plants of January. Thus, although nearly all of the year's plant-life is crowded into February, March, and April, the flower circle around the Twenty Hill Hollow is never broken.

The Hollow may easily be visited by tourists *en route* for Yosemite, as it is distant only about six miles from Snelling's. It is at all seasons interesting to the naturalist; but it has little that would interest the majority of tourists earlier than January or later than April. If you wish to see how much of light, life, and joy can be got into a January, go to this blessed Hollow. If you wish to see a plant-resurrection,—myriads of bright flowers crowding from the ground, like souls to a judgment,—go to Twenty Hills in February. If you are traveling for health, play truant to doctors and friends, fill your pocket with biscuits, and hide in the hills of the Hollow, lave in its waters, tan in its golds, bask in its flower-shine, and your baptisms will make you a new creature indeed. Or, choked in the sediments of society, so tired of the world, here will your hard doubts disappear, your carnal incrustations melt off, and your soul breathe deep and free in God's shoreless atmosphere of beauty and love.

Never shall I forget my baptism in this font. It happened in January, a resurrection day for many a plant and for me. I suddenly found myself on one of its hills; the Hollow overflowed with light, as a fountain, and only small, sunless nooks were kept for mosseries and ferneries. Hollow Creek spangled and mazed like a river. The ground steamed with fragrance. Light, of unspeakable richness, was brooding the flowers. Truly, said I, is California the Golden State — in metallic gold, in sun gold, and in plant gold. The sunshine for a whole summer seemed condensed into the chambers of that one glowing day. Every trace of dimness had been washed from the sky; the mountains were dusted and wiped clean with clouds — Pacheco Peak and Mount Diablo, and the waved blue wall between; the grand Sierra stood along the plain, colored in four horizontal bands: — the lowest, rose purple; the next higher, dark purple; the next, blue; and, above all, the white row of summits pointing to the heavens.

It may be asked, What have mountains fifty or a hundred miles away to do with Twenty Hill Hollow? To lovers of the wild, these mountains are not a hundred miles away. Their spiritual power and the goodness of the sky make them near, as a circle of friends. They rise as a portion of the hilled walls of the Hollow. You cannot feel yourself out of doors; plain, sky, and mountains ray beauty which you feel. You bathe in these spirit-beams, turning round and round, as if warming at a camp-fire. Presently you lose consciousness of your own separate existence: you blend with the landscape, and become part and parcel of nature.

THE SNow

THE first snow that whitens the Sierra usually falls about the end of October or early in
November, to a depth of a few inches, after
months of the most charming Indian summer
weather imaginable. But in a few days, this
light covering mostly melts from the slopes exposed to the sun and causes but little apprehen-
sion on the part of mountaineers who may be
lingering among the high peaks at this time.
The first general winter storm that yields snow
that is to form a lasting portion of the season's
supply, seldom breaks on the mountains before
the end of November. Then, warned by the
sky, cautious mountaineers, together with the
wild sheep, deer, and most of the birds and
bears, make haste to the lowlands or foothills;
and burrowing marmots, mountain beavers,
wood-rats, and such people go into winter quarters, some of them not again to see the light of
day until the general awakening and resurrection of the spring in June or July. The first
heavy fall is usually from about two to four
feet in depth. Then, with intervals of splendid
sunshine, storm succeeds storm, heaping snow

on snow, until thirty to fifty feet has fallen. But on account of its settling and compacting, and the almost constant waste from melting and evaporation, the average depth actually found at any time seldom exceeds ten feet in the forest region, or fifteen feet along the slopes of the summit peaks.

Even during the coldest weather evaporation never wholly ceases, and the sunshine that abounds between the storms is sufficiently powerful to melt the surface more or less through all the winter months. Waste from melting also goes on to some extent on the bottom from heat stored up in the rocks, and given off slowly to the snow in contact with them, as is shown by the rising of the streams on all the higher regions after the first snowfall, and their steady, sustained flow all winter.

The greater portion of the snow deposited around the lofty summits of the range falls in small crisp flakes and broken crystals, or, when accompanied by strong winds and low temperature, the crystals, instead of being locked together in their fall to form tufted flakes, are beaten and broken into meal and fine dust. But down in the forest region the greater portion comes gently to the ground, light and feathery, some of the flakes in mild weather being nearly an inch in diameter, and it is evenly distributed and kept from drifting to any great extent by the shelter afforded by the large trees. Every tree during the progress of gentle storms is loaded with fairy bloom at the coldest and darkest time of year, bending the branches, and hushing every singing needle. But as soon as the storm is over, and the sun shines, the snow at once begins to shift and settle and fall from the branches in miniature avalanches, and the white forest soon becomes green again. The snow on the ground also settles and thaws every bright day, and freezes at night, until it becomes coarsely granulated, and loses every trace of its rayed crystalline structure, and then a man may walk firmly over its frozen surface as if on ice. The forest region up to an elevation of seven thousand feet is usually in great part free from snow in June, but at this time the higher regions are still heavy-laden, and are not touched by spring weather to any considerable extent before the middle or end of July.

One of the most striking effects of the snow on the mountains is the burial of the rivers and small lakes.

> "As the snaw fa's in the river
> A moment white, then lost forever,"—

sang Burns, in illustrating the fleeting character of human pleasure. The first snowflakes that fall into the Sierra rivers vanish thus suddenly; but in great storms, when the temperature is low, the abundance of the snow at length chills the water nearly to the freezing-point, and then, of course, it ceases to melt and consume the snow so suddenly. The falling flakes and crystals form cloud-like masses of blue sludge, which are swept forward with the current and carried down to warmer climates many miles distant, while some are lodged against logs and rocks and projecting points of the banks, and last for days, piled high above the level of the water, and show white again, instead of being at once "lost forever," while the rivers themselves are at length lost for months during the snowy period. The snow is first built out from the banks in bossy, overcurling drifts, compacting and cementing until the streams are spanned. They then flow in the dark beneath a continuous covering across the snowy zone, which is about thirty miles wide. All the Sierra rivers and their tributaries in these high regions are thus lost every winter, as if another glacial period had come on. Not a drop of running water is to be seen excepting at a few points where large falls occur, though the rush and rumble of the heavier currents may still be heard. Toward spring, when the weather is warm during the day and frosty at night, repeated thawing and freezing and new layers of snow render the bridging-masses dense and firm, so that one may safely walk across the streams, or even lead a horse across them without danger of falling through. In June the thinnest parts of the winter ceiling, and those most exposed to sunshine, begin to give way, forming dark, rugged-edged, pit-like sinks, at the bottom of which the rushing water may be seen. At the end of June only here and there may the mountaineer find a secure snow-bridge. The most lasting of the winter bridges, thawing from below as well as from above, because of warm currents of air passing through the tunnels, are strikingly arched and sculptured; and by the occasional freezing of the oozing, dripping water of the ceiling they become brightly and picturesquely icy. In some of the reaches, where there is a free margin, we may walk through them. Small skylights appearing here and there, these tunnels are not very dark. The roaring river fills all the arching way with impressively loud reverberating music, which is sweetened at times by the ouzel, a bird that is not afraid to go wherever a stream may go, and to sing wherever a stream sings.

All the small alpine pools and lakelets are in like manner obliterated from the winter landscapes, either by being first frozen and then covered by snow, or by being filled in by avalanches. The first avalanche of the season shot into a lake basin may perhaps find the surface frozen. Then there is a grand crashing of breaking ice and dashing of waves mingled with the low, deep booming of the avalanche. Detached masses of the invading snow, mixed with fragments of ice, drift about in sludgy, island-like heaps, while the main body of it forms a talus with its base wholly or in part resting on the bottom of the basin, as controlled by its depth and the size of the avalanche. The next avalanche, of course, encroaches still farther, and so on with each in succession until the entire basin may be filled and its water sponged up or displaced. This huge mass of sludge, more or less mixed with sand, stones, and perhaps timber, is frozen to a considerable depth, and much sun-heat is required to thaw it. Some of these unfortunate lakelets are not clear of ice and snow until near the end of summer. Others are never quite free, opening only on the side opposite the entrance of the avalanches. Some show only a narrow crescent of water lying between the shore and sheer bluffs of icy compacted snow, masses of which breaking off float in front like icebergs in a miniature Arctic Ocean, while the avalanche heaps leaning back against the mountains look like small glaciers. The frontal cliffs are in some instances quite picturesque, and with the berg-dotted waters in front of them lighted with sunshine are exceedingly beautiful. It often happens that while one side of a lake basin is hopelessly snow-buried and frozen, the other, enjoying sunshine, is adorned with beautiful flower-gardens. Some of the smaller lakes are extinguished in an instant by a heavy avalanche either of rocks or snow. The rolling, sliding, ponderous mass entering on one side sweeps across the bottom and up the opposite side, displacing the water and even scraping the basin clean, and shoving the accumulated rocks and sediments up the farther bank and taking full possession. The dislodged water is in part absorbed, but most of it is sent around the front of the avalanche and down the channel of the outlet, roaring and hurrying as if frightened and glad to escape.

SNOW-BANNERS

The most magnificent storm phenomenon I ever saw, surpassing in showy grandeur the most imposing effects of clouds, floods, or avalanches, was the peaks of the High Sierra, back of Yosemite Valley, decorated with snow-banners. Many of the starry snow-flowers, out of which these banners are made, fall before they are ripe, while most of those that do attain perfect development as six-rayed crystals glint and chafe against one another in their fall through the frosty air, and are broken into fragments. This dry fragmentary snow is still further prepared for the formation of banners by the action of the wind. For, instead of finding rest at once, like the snow which falls into the tranquil depths of the forests, it is rolled over and over, beaten against rock ridges, and swirled in pits and hollows, like boulders, pebbles, and sand in the pot-holes of a river, until finally the delicate angles of the crystals are worn off, and the whole mass is reduced to dust. And whenever storm-winds find this prepared snow-dust in a loose condition on exposed slopes, where there is a free upward sweep to leeward, it is tossed back into the sky, and borne onward from peak to peak in the form of banners or cloudy drifts, according to the velocity of the wind and the conformation of the slopes up or around which it is driven. While thus flying through the air, a small portion makes good its escape, and remains in the sky as vapor. But far the greater part, after being driven into the sky again and again, is at length locked fast in bossy drifts, or in the wombs of glaciers, some of it to remain silent and rigid for centuries before it is finally melted and sent singing down the mountain-sides to the sea.

Yet, notwithstanding the abundance of winter snow-dust in the mountains, and the frequency of high winds, and the length of time the dust remains loose and exposed to their action, the occurrence of well-formed banners is, for causes we shall hereafter note, comparatively rare. I have seen only one display of this kind that seemed in every way perfect. This was in the winter of 1873, when the snow-laden summits were swept by a wild "norther." I happened at the time to be wintering in Yosemite Valley, that sublime Sierra temple where every day one may see the grandest sights. Yet even here the wild gala day of the north wind seemed surpassingly glorious. I was awakened in the morning by the rocking of my cabin and the beating of pine burs on the roof. Detached torrents and avalanches from the main wind flood overhead were rushing wildly down the narrow side cañons, and over the precipitous walls, with loud resounding roar, rousing the pines to enthusiastic action, and making the whole valley vibrate as though it were an instrument being played.

But afar on the lofty exposed peaks of the range standing so high in the sky, the storm was expressing itself in still grander characters, which I was soon to see in all their glory. I had long been anxious to study some points in the structure of the ice cone that is formed every winter at the foot of the Upper Yosemite Fall, but the blinding spray by which it is invested had hitherto prevented me from making a sufficiently near approach. This morning the entire body of the fall was torn into gauzy shreds, and blown horizontally along the face of the cliff, leaving the cone dry; and while making my way to the top of an overlooking ledge to seize so favorable an opportunity to examine the interior of the cone, the peaks of the Merced group came in sight over the shoulder of the South Dome, each waving a resplendent banner against the blue sky, as regular in form, and as firm in texture, as if woven of fine silk. So rare and splendid a phenomenon, of course, overbore all other considerations, and I at once let the ice cone go, and began to force my way out of the valley to some dome or ridge sufficiently lofty to command a general view of the main summits, feeling assured that I should find them bannered still more gloriously; nor was I in the least disappointed. Indian Cañon, through which I climbed, was choked with snow that had been shot down in avalanches from the high cliffs on either side, rendering the ascent

THE SNOW

difficult; but inspired by the roaring storm, the
tedious wallowing brought no fatigue, and in
four hours I gained the top of a ridge above
the valley, eight thousand feet high. And there
in bold relief, like a clear painting, appeared
a most imposing scene. Innumerable peaks,
black and sharp, rose grandly into the dark
blue sky, their bases set in solid white, their
sides streaked and splashed with snow, like
ocean rocks with foam; and from every summit, all free and unconfused, was streaming
a beautiful silky silvery banner, from half a
mile to a mile in length, slender at the point
of attachment, then widening gradually as it
extended from the peak until it was about one
thousand or fifteen hundred feet in breadth, as
near as I could estimate. The cluster of peaks
called the "Crown of the Sierra," at the head
of the Merced and Tuolumne Rivers, — Mounts
Dana, Gibbs, Conness, Lyell, Maclure, Ritter,
with their nameless compeers, — each had its
own refulgent banner, waving with a clearly
visible motion in the sunglow, and there was
not a single cloud in the sky to mar their simple
grandeur. Fancy yourself standing on this
Yosemite ridge looking eastward. You notice a
strange garish glitter in the air. The gale drives
wildly overhead with a fierce, tempestuous roar,
but its violence is not felt, for you are looking

through a sheltered opening in the woods as through a window. There, in the immediate foreground of your picture, rises a majestic forest of silver fir blooming in eternal freshness, the foliage yellow-green, and the snow beneath the trees strewn with their beautiful plumes, plucked off by the wind. Beyond, and extending over all the middle ground, are somber swaths of pine, interrupted by huge swelling ridges and domes; and just beyond the dark forest you see the monarchs of the High Sierra waving their magnificent banners. They are twenty miles away, but you would not wish them nearer, for every feature is distinct, and the whole glorious show is seen in its right proportions. After this general view, mark how sharply the dark snowless ribs and buttresses and summits of the peaks are defined, excepting the portions veiled by the banners, and how delicately their sides are streaked with snow, where it has come to rest in narrow flutings and gorges. Mark, too, how grandly the banners wave as the wind is deflected against their sides, and how trimly each is attached to the very summit of its peak, like a streamer at a masthead; how smooth and silky they are in texture, and how finely their fading fringes are penciled on the azure sky. See how dense and opaque they are at the point of attachment,

and how filmy and translucent toward the end,
so that the peaks back of them are seen dimly,
as though you were looking through ground
glass. Yet again observe how some of the longest, belonging to the loftiest summits, stream
perfectly free all the way across intervening
notches and passes from peak to peak, while
others overlap and partly hide each other. And
consider how keenly every particle of this wondrous cloth of snow is flashing out jets of light.
These are the main features of the beautiful
and terrible picture as seen from the forest
window; and it would still be surpassingly glorious were the fore- and middle-grounds obliterated altogether, leaving only the black peaks,
the white banners, and the blue sky.

Glancing now in a general way at the formation of snow-banners, we find that the main
causes of the wondrous beauty and perfection
of those we have been contemplating were the
favorable direction and great force of the wind,
the abundance of snow-dust, and the peculiar
conformation of the slopes of the peaks. It is
essential not only that the wind should move
with great velocity and steadiness to supply a
sufficiently copious and continuous stream of
snow-dust, but that it should come from the
north. No perfect banner is ever hung on the
Sierra peaks by a south wind. Had the gale

that day blown from the south, leaving other conditions unchanged, only a dull, confused, foglike drift would have been produced; for the snow, instead of being spouted up over the tops of the peaks in concentrated currents to be drawn out as streamers, would have been shed off around the sides, and piled down into the glacier wombs. The cause of the concentrated action of the north wind is found in the peculiar form of the north sides of the peaks, where the amphitheaters of the residual glaciers are. In general the south sides are convex and irregular, while the north sides are concave in both their vertical and horizontal sections; the wind in ascending these curves converges toward the summits, carrying the snow in concentrating currents with it, shooting it almost straight up into the air above the peaks, from which it is then carried away in a horizontal direction.

This difference in form between the north and south sides of the peaks was almost wholly produced by the difference in the kind and quantity of the glaciation to which they have been subjected, the north sides having been hollowed by residual shadow glaciers of a form that never existed on the sun-beaten sides.

It appears, therefore, that shadows in great part determine not only the forms of lofty icy mountains, but also those of the snow-banners that the wild winds hang on them.

A NEAR VIEW OF THE HIGH SIERRA

EARLY one bright morning in the middle of Indian summer, while the glacier meadows were still crisp with frost crystals, I set out from the foot of Mount Lyell, on my way down to Yosemite Valley, to replenish my exhausted store of bread and tea. I had spent the past summer, as many preceding ones, exploring the glaciers that lie on the head waters of the San Joaquin, Tuolumne, Merced, and Owen's Rivers; measuring and studying their movements, trends, crevasses, moraines, etc., and the part they had played during the period of their greater extension in the creation and development of the landscapes of this alpine wonderland. The time for this kind of work was nearly over for the year, and I began to look forward with delight to the approaching winter with its wondrous storms, when I would be warmly snow-bound in my Yosemite cabin with plenty of bread and books; but a tinge of regret came on when I considered that possibly I might not see this favorite region again until the next summer, excepting distant views from the heights about the Yosemite walls.

The Mountains of California, 1894

To artists, few portions of the High Sierra are, strictly speaking, picturesque. The whole massive uplift of the range is one great picture, not clearly divisible into smaller ones; differing much in this respect from the older, and what may be called, riper mountains of the Coast Range. All the landscapes of the Sierra, as we have seen, were born again, remodeled from base to summit by the developing ice floods of the last glacial winter. But all these new landscapes were not brought forth simultaneously; some of the highest, where the ice lingered longest, are tens of centuries younger than those of the warmer regions below them. In general, the younger the mountain landscapes, — younger, I mean, with reference to the time of their emergence from the ice of the glacial period, — the less separable are they into artistic bits capable of being made into warm, sympathetic, lovable pictures with appreciable humanity in them.

Here, however, on the head waters of the Tuolumne, is a group of wild peaks on which the geologist may say that the sun has but just begun to shine, which is yet in a high degree picturesque, and in its main features so regular and evenly balanced as almost to appear conventional — one somber cluster of snow-laden peaks with gray, pine-fringed, granite bosses

braided around its base, the whole surging free
into the sky from the head of a magnificent
valley, whose lofty walls are beveled away on
both sides so as to embrace it all without ad-
mitting anything not strictly belonging to it.
The foreground was now aflame with autumn
colors, brown and purple and gold, ripe in the
mellow sunshine; contrasting brightly with the
deep, cobalt blue of the sky, and the black and
gray, and pure, spiritual white of the rocks and
glaciers. Down through the midst, the young
Tuolumne was seen pouring from its crystal
fountains, now resting in glassy pools as if
changing back again into ice, now leaping in
white cascades as if turning to snow; gliding
right and left between granite bosses, then
sweeping on through the smooth, meadowy
levels of the valley, swaying pensively from
side to side with calm, stately gestures past
dipping willows and sedges, and around groves
of arrowy pine; and throughout its whole eventful course, whether flowing fast or slow, singing
loud or low, ever filling the landscape with
spiritual animation, and manifesting the grandeur of its sources in every movement and
tone.

Pursuing my lonely way down the valley, I
turned again and again to gaze on the glorious
picture, throwing up my arms to inclose it as in

a frame. After long ages of growth in the darkness beneath the glaciers, through sunshine and storms, it seemed now to be ready and waiting for the elected artist, like yellow wheat for the reaper; and I could not help wishing that I might carry colors and brushes with me on my travels, and learn to paint. In the mean time I had to be content with photographs on my mind and sketches in my notebooks. At length, after I had rounded a precipitous headland that puts out from the west wall of the valley, every peak vanished from sight, and I pushed rapidly along the frozen meadows, over the divide between the waters of the Merced and Tuolumne, and down through the forests that clothe the slopes of Cloud's Rest, arriving in Yosemite in due time — which, with me, is *any* time. And, strange to say, among the first people I met here were two artists who, with letters of introduction, were awaiting my return. They inquired whether in the course of my explorations in the adjacent mountains I had ever come upon a landscape suitable for a large painting; whereupon I began a description of the one that had so lately excited my admiration. Then, as I went on further and further into details, their faces began to glow, and I offered to guide them to it, while they declared that they would gladly follow, far or near, whithersoever I could spare the time to lead them.

Since storms might come breaking down through the fine weather at any time, burying the colors in snow, and cutting off the artists' retreat, I advised getting ready at once.

I led them out of the valley by the Vernal and Nevada Falls, thence over the main dividing ridge to the Big Tuolumne Meadows, by the old Mono Trail, and thence along the Upper Tuolumne River to its head. This was my companions' first excursion into the High Sierra, and as I was almost always alone in my mountaineering, the way that the fresh beauty was reflected in their faces made for me a novel and interesting study. They naturally were affected most of all by the colors — the intense azure of the sky, the purplish grays of the granite, the red and browns of dry meadows, and the translucent purple and crimson of huckleberry bogs; the flaming yellow of aspen groves, the silvery flashing of the streams, and the bright green and blue of the glacier lakes. But the general expression of the scenery — rocky and savage — seemed sadly disappointing; and as they threaded the forest from ridge to ridge, eagerly scanning the landscapes as they were unfolded, they said: "All this is huge and sublime, but we see nothing as yet at all available for effective pictures. Art is long, and art is limited, you know; and here are foregrounds, middle-grounds, backgrounds, all alike; bare rock waves, woods, groves, diminutive flecks of meadow, and strips of glittering water." "Never mind," I replied, "only bide a wee, and I will show you something you will like."

At length, toward the end of the second day, the Sierra Crown began to come into view, and when we had fairly rounded the projecting headland before mentioned, the whole picture stood revealed in the flush of the alpenglow. Their enthusiasm was excited beyond bounds, and the more impulsive of the two, a young Scotchman, dashed ahead, shouting and gesticulating and tossing his arms in the air like a madman. Here, at last, was a typical alpine landscape.

After feasting a while on the view, I proceeded to make camp in a sheltered grove a little way back from the meadow, where pine boughs could be obtained for beds, and where there was plenty of dry wood for fires, while the artists ran here and there, along the river bends and up the sides of the cañon, choosing foregrounds for sketches. After dark, when our tea was made and a rousing fire had been built, we began to make our plans. They decided to remain several days, at the least, while I concluded to make an excursion in the mean time to the untouched summit of Ritter.

It was now about the middle of October, the springtime of snow-flowers. The first winter clouds had already bloomed, and the peaks were strewn with fresh crystals, without, however, affecting the climbing to any dangerous extent. And as the weather was still profoundly calm, and the distance to the foot of the mountain only a little more than a day, I felt that I was running no great risk of being storm-bound.

Mount Ritter is king of the mountains of the middle portion of the High Sierra, as Shasta of the north and Whitney of the south sections. Moreover, as far as I know, it had never been climbed. I had explored the adjacent wilderness summer after summer, but my studies thus far had never drawn me to the top of it. Its height above sea-level is about 13,300 feet, and it is fenced round by steeply inclined glaciers, and cañons of tremendous depth and ruggedness, which render it almost inaccessible. But difficulties of this kind only exhilarate the mountaineer.

Next morning, the artists went heartily to their work and I to mine. Former experiences had given good reason to know that passionate storms, invisible as yet, might be brooding in the calm sungold; therefore, before bidding farewell, I warned the artists not to be alarmed should I fail to appear before a week or ten days, and advised them, in case a snowstorm should set in, to keep up big fires and shelter themselves as best they could, and on no account to become frightened and attempt to seek their way back to Yosemite alone through the drifts.

My general plan was simply this: to scale the cañon wall, cross over to the eastern flank of the range, and then make my way southward to the northern spurs of Mount Ritter in compliance with the intervening topography; for to push on directly southward from camp through the innumerable peaks and pinnacles that adorn this portion of the axis of the range, however interesting, would take too much time, besides being extremely difficult and dangerous at this time of year.

All my first day was pure pleasure; simply mountaineering indulgence, crossing the dry pathways of the ancient glaciers, tracing happy streams, and learning the habits of the birds and marmots in the groves and rocks. Before I had gone a mile from camp, I came to the foot of a white cascade that beats its way down a rugged gorge in the cañon wall, from a height of about nine hundred feet, and pours its throbbing waters into the Tuolumne. I was acquainted with its fountains, which, fortunately, lay in my course. What a fine traveling companion it proved to be, what songs it sang, and how passionately it told the mountain's own joy! Gladly I climbed along its dashing border, absorbing its divine music, and bathing from time to time in waftings of irised spray. Climbing higher, higher, new beauty came streaming on the sight: painted meadows, late-blooming gardens, peaks of rare architecture, lakes here and there, shining like silver, and glimpses of the forested middle region and the yellow lowlands far in the west. Beyond the range I saw the so-called Mono Desert, lying dreamily silent in thick purple light — a desert of heavy sunglare beheld from a desert of ice-burnished granite. Here the waters divide, shouting in glorious enthusiasm, and falling eastward to vanish in the volcanic sands and dry sky of the Great Basin, or westward to the Great Valley of California, and thence through the Bay of San Francisco and the Golden Gate to the sea.

Passing a little way down over the summit until I had reached an elevation of about ten thousand feet, I pushed on southward toward a group of savage peaks that stand guard about Ritter on the north and west, groping my way, and dealing instinctively with every obstacle as it presented itself. Here a huge gorge would be found cutting across my path, along the dizzy edge of which I scrambled until some less precipitous point was discovered where I might safely venture to the bottom and then, selecting some feasible portion of the opposite wall, reascend with the same slow caution. Massive, flat-topped spurs alternate with the gorges, plunging abruptly from the shoulders of the snowy peaks, and planting their feet in the warm desert. These were everywhere marked and adorned with characteristic sculptures of the ancient glaciers that swept over this entire region like one vast ice wind, and the polished surfaces produced by the ponderous flood are still so perfectly preserved that in many places the sunlight reflected from them is about as trying to the eyes as sheets of snow.

God's glacial mills grind slowly, but they have been kept in motion long enough in California to grind sufficient soil for a glorious abundance of life, though most of the grist has been carried to the lowlands, leaving these high regions comparatively lean and bare; while the post-glacial agents of erosion have not yet furnished sufficient available food over the general surface for more than a few tufts of the hardiest plants, chiefly carices and eriogonæ. And it is interesting to learn in this connection that the sparseness and repressed character of the vegetation at this height is caused more by want of soil than by harshness of climate; for, here and there, in sheltered hollows (countersunk beneath the general surface) into which a few rods of well-ground moraine chips have been dumped, we find groves of spruce and pine thirty to forty feet high, trimmed around the edges with willow and huckleberry bushes, and oftentimes still further by an outer ring of tall grasses, bright with lupines, larkspurs, and showy columbines, suggesting a climate by no means repressingly severe. All the streams, too, and the pools at this elevation are furnished with little gardens wherever soil can be made to lie, which, though making scarce any show at a distance, constitute charming surprises to the appreciative observer. In these bits of leafiness a few birds find grateful homes. Having no acquaintance with man, they fear no ill, and flock curiously about the stranger, almost allowing themselves to be taken in the hand. In so wild and so beautiful a region was spent my first day, every sight and sound inspiring, leading one far out of himself, yet feeding and building up his individuality.

Now came the solemn, silent evening. Long, blue, spiky shadows crept out across the snow-fields, while a rosy glow, at first scarce discernible, gradually deepened and suffused every mountain-top, flushing the glaciers and the harsh crags above them. This was the alpenglow, to me one of the most impressive of all the terrestrial manifestations of God. At the touch of this divine light, the mountains seemed to kindle to a rapt, religious consciousness, and stood hushed and waiting like devout worshipers. Just before the alpenglow began to fade, two crimson clouds came streaming across the summit like wings of flame, rendering the sublime scene yet more impressive; then came darkness and the stars.

Icy Ritter was still miles away, but I could proceed no farther that night. I found a good camp-ground on the rim of a glacier basin about eleven thousand feet above the sea. A small lake nestles in the bottom of it, from which I got water for my tea, and a storm-beaten thicket near by furnished abundance of resiny firewood. Somber peaks, hacked and shattered, circled halfway around the horizon, wearing a savage aspect in the gloaming, and a waterfall chanted solemnly across the lake on its way down from the foot of a glacier. The fall and the lake and the glacier were almost equally bare; while the scraggy pines anchored in the rock-fissures were so dwarfed and shorn by storm-winds that you might walk over their tops. In tone and aspect the scene was one of the most desolate I ever beheld. But the darkest scriptures of the mountains are illumined with bright passages of love that never fail to make themselves felt when one is alone.

I made my bed in a nook of the pine thicket, where the branches were pressed and crinkled overhead like a roof, and bent down around the sides. These are the best bedchambers the high mountains afford — snug as squirrel nests, well-ventilated, full of spicy odors, and with plenty of wind-played needles to sing one asleep. I little expected company, but, creeping in through a low side door, I found five or six birds nestling among the tassels. The night wind began to blow soon after dark; at first only a gentle breathing, but increasing toward midnight to a rough gale that fell upon my leafy roof in ragged surges like a cascade, bearing wild sounds from the crags overhead. The waterfall sang in chorus, filling the old ice fountain with its solemn roar, and seeming to increase in power as the night advanced — fit voice for such a landscape. I had to creep out many times to the fire during the night, for it was biting cold and I had no blankets. Gladly I welcomed the morning star.

The dawn in the dry, wavering air of the desert was glorious. Everything encouraged my undertaking and betokened success. There was no cloud in the sky, no storm tone in the wind. Breakfast of bread and tea was soon made. I fastened a hard, durable crust to my belt by way of provision, in case I should be compelled to pass a night on the mountain-top; then, securing the remainder of my little stock against wolves and wood rats, I set forth free and hopeful.

How glorious a greeting the sun gives the mountains! To behold this alone is worth the pains of any excursion a thousand times over. The highest peaks burned like islands in a sea of liquid shade. Then the lower peaks and spires caught the glow, and long lances of light, streaming through many a notch and pass, fell thick on the frozen meadows. The majestic form of Ritter was full in sight, and I pushed rapidly on over rounded rock bosses and pave-ments, my iron-shod shoes making a clanking sound, suddenly hushed now and then in rugs of bryanthus, and sedgy lake margins soft as moss. Here, too, in this so-called "land of des-olation," I met cassiope, growing in fringes among the battered rocks. Her blossoms had faded long ago, but they were still clinging with happy memories to the evergreen sprays, and still so beautiful as to thrill every fiber of one's being. Winter and summer, you may hear her voice, the low, sweet melody of her purple bells.

A NEAR VIEW OF THE HIGH SIERRA

No evangel among all the mountain plants speaks Nature's love more plainly than cassiope. Where she dwells, the redemption of the coldest solitude is complete. The very rocks and glaciers seem to feel her presence, and become imbued with her own fountain sweetness. All things were warming and awakening. Frozen rills began to flow, the marmots came out of their nests in boulder piles and climbed sunny rocks to bask, and the dun-headed sparrows were flitting about seeking their breakfasts. The lakes seen from every ridge-top were brilliantly rippled and spangled, shimmering like the thickets of the low dwarf pines. The rocks, too, seemed responsive to the vital heat — rock crystals and snow crystals thrilling alike. I strode on exhilarated, as if never more to feel fatigue, limbs moving of themselves, every sense unfolding like the thawing flowers, to take part in the new day harmony.

All along my course thus far, excepting when down in the cañons, the landscapes were mostly open to me, and expansive, at least on one side. On the left were the purple plains of Mono, reposing dreamily and warm; on the right, the near peaks springing keenly into the thin sky with more and more impressive sublimity. But these larger views were at length lost. Rugged spurs, and moraines, and huge, projecting

buttresses began to shut me in. Every feature
became more rigidly alpine, without, however,
producing any chilling effect; for going to the
mountains is like going home. We always find
that the strangest objects in these fountain
wilds are in some degree familiar, and we look
upon them with a vague sense of having seen
them before.

On the southern shore of a frozen lake, I encountered an extensive field of hard, granular
snow, up which I scampered in fine tone, intending to follow it to its head, and cross the
rocky spur against which it leans, hoping thus
to come direct upon the base of the main Ritter
peak. The surface was pitted with oval hollows, made by stones and drifted pine needles
that had melted themselves into the mass by
the radiation of absorbed sun-heat. These
afforded good footholds, but the surface curved
more and more steeply at the head, and the
pits became shallower and less abundant, until
I found myself in danger of being shed off like
avalanching snow. I persisted, however, creeping on all fours, and shuffling up the smoothest
places on my back, as I had often done on
burnished granite, until, after slipping several
times, I was compelled to retrace my course to
the bottom, and made my way around the
west end of the lake, and thence up to the sum-

mit of the divide between the head waters of
Rush Creek and the northernmost tributaries
of the San Joaquin.

Arriving on the summit of this dividing crest,
one of the most exciting pieces of pure wilderness was disclosed that I ever discovered in
all my mountaineering. There, immediately
in front, loomed the majestic mass of Mount
Ritter, with a glacier swooping down its face
nearly to my feet, then curving westward and
pouring its frozen flood into a dark blue lake,
whose shores were bound with precipices of
crystalline snow; while a deep chasm drawn
between the divide and the glacier separated
the massive picture from everything else. I
could see only the one sublime mountain, the
one glacier, the one lake; the whole veiled with
one blue shadow — rock, ice, and water close
together, without a single leaf or sign of life.
After gazing spellbound, I began instinctively
to scrutinize every notch and gorge and weathered buttress of the mountain, with reference
to making the ascent. The entire front above
the glacier appeared as one tremendous precipice, slightly receding at the top, and bristling
with spires and pinnacles set above one another
in formidable array. Massive lichen-stained
battlements stood forward here and there,
hacked at the top with angular notches, and

separated by frosty gullies and recesses that have been veiled in shadow ever since their creation; while to right and left, as far as I could see, were huge, crumbling buttresses, offering no hope to the climber. The head of the glacier sends up a few finger-like branches through narrow *couloirs;* but these seemed too steep and short to be available, especially as I had no axe with which to cut steps, and the numerous narrow-throated gullies down which stones and snow are avalanched seemed hopelessly steep, besides being interrupted by vertical cliffs; while the whole front was rendered still more terribly forbidding by the chill shadow and the gloomy blackness of the rocks.

Descending the divide in a hesitating mood, I picked my way across the yawning chasm at the foot, and climbed out upon the glacier. There were no meadows now to cheer with their brave colors, nor could I hear the dun-headed sparrows, whose cheery notes so often relieve the silence of our highest mountains. The only sounds were the gurgling of small rills down in the veins and crevasses of the glacier, and now and then the rattling report of falling stones, with the echoes they shot out into the crisp air.

I could not distinctly hope to reach the summit from this side, yet I moved on across the glacier as if driven by fate. Contending with myself, the season is too far spent, I said, and even should I be successful, I might be storm-bound on the mountain; and in the cloud darkness, with the cliffs and crevasses covered with snow, how could I escape? No; I must wait till next summer. I would only approach the mountain now, and inspect it, creep about its flanks, learn what I could of its history, holding myself ready to flee on the approach of the first storm cloud. But we little know until tried how much of the uncontrollable there is in us, urging over glaciers and torrents, and up perilous heights, let the judgment forbid as it may.

I succeeded in gaining the foot of the cliff on the eastern extremity of the glacier, and there discovered the mouth of a narrow avalanche gully, through which I began to climb, intending to follow it as far as possible, and at least obtain some fine wild views for my pains. Its general course is oblique to the plane of the mountain-face, and the metamorphic slates of which the mountain is built are cut by cleavage planes in such a way that they weather off in angular blocks, giving rise to irregular steps that greatly facilitate climbing on the sheer places. I thus made my way into a wilderness of crumbling spires and battlements, built together in bewildering combinations, and glazed in many places with a thin coating of ice, which I had to hammer off with stones. The situation was becoming gradually more perilous; but, having passed several dangerous spots, I dared not think of descending; for, so steep was the entire ascent, one would inevitably fall to the glacier in case a single misstep were made. Knowing, therefore, the tried danger beneath, I became all the more anxious concerning the developments to be made above, and began to be conscious of a vague foreboding of what actually befell; not that I was given to fear, but rather because my instincts, usually so positive and true, seemed vitiated in some way, and were leading me astray. At length, after attaining an elevation of about 12,800 feet, I found myself at the foot of a sheer drop in the bed of the avalanche channel I was tracing, which seemed absolutely to bar further progress. It was only about forty-five or fifty feet high, and somewhat roughened by fissures and projections; but these seemed so slight and insecure, as footholds, that I tried hard to avoid the precipice altogether, by scaling the wall of the channel on either side. But, though less steep, the walls were smoother than the obstructing rock, and repeated efforts only showed that I must either go right ahead or turn back. The tried dangers beneath seemed even greater than that of the cliff in front; therefore, after scanning its face again and again, I began to scale it, picking my holds with intense caution. After gaining a point about halfway to the top, I was suddenly brought to a dead stop, with arms outspread, clinging close to the face of the rock, unable to move hand or foot either up or down. My doom appeared fixed. I *must* fall. There would be a moment of bewilderment, and then a lifeless rumble down the one general precipice to the glacier below.

When this final danger flashed upon me, I became nerve-shaken for the first time since setting foot on the mountains, and my mind seemed to fill with a stifling smoke. But this terrible eclipse lasted only a moment, when life blazed forth again with preternatural clearness. I seemed suddenly to become possessed of a new sense. The other self, bygone experiences, Instinct, or Guardian Angel, — call it what you will, — came forward and assumed control. Then my trembling muscles became firm again, every rift and flaw in the rock was seen as through a microscope, and my limbs moved with a positiveness and precision with which I seemed to have nothing at all to do. Had I been borne aloft upon wings, my deliverance could not have been more complete.

Above this memorable spot, the face of the mountain is still more savagely hacked and torn. It is a maze of yawning chasms and gullies, in the angles of which rise beetling crags and piles of detached boulders that seem to have been gotten ready to be launched below. But the strange influx of strength I had received seemed inexhaustible. I found a way without effort, and soon stood upon the topmost crag in the blessed light.

How truly glorious the landscape circled around this noble summit! — giant mountains, valleys innumerable, glaciers and meadows, rivers and lakes, with the wide blue sky bent tenderly over them all. But in my first hour of freedom from that terrible shadow, the sunlight in which I was laving seemed all in all.

Looking southward along the axis of the range, the eye is first caught by a row of exceedingly sharp and slender spires, which rise openly to a height of about a thousand feet, above a series of short, residual glaciers that lean back against their bases; their fantastic sculpture and the unrelieved sharpness with which they spring out of the ice rendering them peculiarly wild and striking. These are "The Minarets." Beyond them you behold a sublime wilderness of mountains, their snowy summits towering together in crowded abundance, peak beyond peak, swelling higher, higher, as they sweep on southward, until the culminating point of the range is reached on Mount Whitney, near the head of the Kern River, at an elevation of nearly 14,700 feet above the level of the sea.

Westward, the general flank of the range is seen flowing sublimely away from the sharp summits, in smooth undulations; a sea of huge gray granite waves dotted with lakes and meadows, and fluted with stupendous cañons that grow steadily deeper as they recede in the distance. Below this gray region lies the dark forest zone, broken here and there by upswelling ridges and domes; and yet beyond lies a yellow, hazy belt, marking the broad plain of the San Joaquin, bounded on its farther side by the blue mountains of the coast.

Turning now to the northward, there in the immediate foreground is the glorious Sierra Crown, with Cathedral Peak, a temple of marvelous architecture, a few degrees to the left of it; the gray, massive form of Mammoth Mountain to the right; while Mounts Ord, Gibbs, Dana, Conness, Tower Peak, Castle Peak, Silver Mountain, and a host of noble companions, as yet nameless, make a sublime show along the axis of the range.

Eastward, the whole region seems a land of

desolation covered with beautiful light. The torrid volcanic basin of Mono, with its one bare lake fourteen miles long; Owen's Valley and the broad lava tableland at its head, dotted with craters, and the massive Inyo Range, rivaling even the Sierra in height; these are spread, map-like, beneath you, with countless ranges beyond, passing and overlapping one another and fading on the glowing horizon.

At a distance of less than three thousand feet below the summit of Mount Ritter you may find tributaries of the San Joaquin and Owen's Rivers, bursting forth from the ice and snow of the glaciers that load its flanks; while a little to the north of here are found the highest affluents of the Tuolumne and Merced. Thus, the fountains of four of the principal rivers of California are within a radius of four or five miles.

Lakes are seen gleaming in all sorts of places, — round, or oval, or square, like very mirrors; others narrow and sinuous, drawn close around the peaks like silver zones, the highest reflecting only rocks, snow, and the sky. But neither these nor the glaciers, nor the bits of brown meadow and moorland that occur here and there, are large enough to make any marked impression upon the mighty wilderness of mountains. The eye, rejoicing in its freedom, roves about the vast expanse, yet returns again and again to the fountain peaks. Perhaps some one of the multitude excites special attention, some gigantic castle with turret and battlement, or some Gothic cathedral more abundantly spired than Milan's. But, generally, when looking for the first time from an all-embracing standpoint like this, the inexperienced observer is oppressed by the incomprehensible grandeur, variety, and abundance of the mountains rising shoulder to shoulder beyond the reach of vision; and it is only after they have been studied one by one, long and lovingly, that their far-reaching harmonies become manifest. Then, penetrate the wilderness where you may, the main telling features, to which all the surrounding topography is subordinate, are quickly perceived, and the most complicated clusters of peaks stand revealed harmoniously correlated and fashioned like works of art — eloquent monuments of the ancient ice rivers that brought them into relief from the general mass of the range. The cañons, too, some of them a mile deep, mazing wildly through the mighty host of mountains, however lawless and ungovernable at first sight they appear, are at length recognized as the necessary effects of causes which followed each other in harmonious sequence—Nature's poems carved on tables of stone — simplest and most emphatic of her glacial compositions.

Could we have been here to observe during the glacial period, we should have overlooked a wrinkled ocean of ice as continuous as that now covering the landscapes of Greenland; filling every valley and cañon with only the tops of the fountain peaks rising darkly above the rock-encumbered ice waves like islets in a stormy sea — those islets the only hints of the glorious landscapes now smiling in the sun. Standing here in the deep, brooding silence all the wilderness seems motionless, as if the work of creation were done. But in the midst of this outer steadfastness we know there is incessant motion and change. Ever and anon, avalanches are falling from yonder peaks. These cliff-bound glaciers, seemingly wedged and immovable, are flowing like water and grinding the rocks beneath them. The lakes are lapping their granite shores and wearing them away, and every one of these rills and young rivers is fretting the air into music, and carrying the mountains to the plains. Here are the roots of all the life of the valleys, and here more simply than elsewhere is the eternal flux of Nature manifested. Ice changing to water, lakes to meadows, and mountains to plains. And while we thus contemplate Nature's methods of landscape creation, and, reading the records she has carved on the rocks, reconstruct, however imperfectly, the landscapes of the past, we also learn that as these we now behold have succeeded those of the pre-glacial age, so they in turn are withering and vanishing to be succeeded by others yet unborn.

But in the midst of these fine lessons and landscapes, I had to remember that the sun was wheeling far to the west, while a new way down the mountain had to be discovered to some point on the timber line where I could have a fire; for I had not even burdened myself with a coat. I first scanned the western spurs, hoping some way might appear through which I might reach the northern glacier, and cross its snout, or pass around the lake into which it flows, and thus strike my morning track. This route was soon sufficiently unfolded to show that, if it were practicable at all, it would require so much time that reaching camp that night would be out of the question. I therefore scrambled back eastward, and descended the southern slopes obliquely at the same time. Here the crags seemed less formidable, and the head of a glacier that flows northeast came in sight, which I determined to follow as far as possible, hoping thus to make my way to the foot of the peak on the east side, and thence

across the intervening cañons and ridges to
camp.

The inclination of the glacier is quite moderate at the head, and, as the sun had softened
the *névé*, I made safe and rapid progress, running and sliding, and keeping up a sharp outlook for crevasses. About half a mile from the
head, there is an ice cascade, where the glacier
pours over a sharp declivity and is shattered
into massive blocks separated by deep, blue
fissures. To thread my way through the slip-
pery mazes of this crevassed portion seemed
impossible, and I endeavored to avoid it by
climbing off to the shoulder of the mountain.
But the slopes rapidly steepened and at length
fell away in sheer precipices, compelling a return to the ice. Fortunately, the day had been
warm enough to loosen the ice crystals so as to
admit of hollows being dug in the rotten portions of the blocks, thus enabling me to pick
my way with far less difficulty than I had anti-
cipated. Continuing down over the snout, and
along the left lateral moraine, was only a confident saunter, showing that the ascent of the
mountain by way of this glacier is easy, provided one is armed with an axe to cut steps
here and there.

The lower end of the glacier was beautifully
waved and barred by the outcropping edges of

the bedded ice layers which represent the annual snowfalls, and to some extent the irregularities of structure caused by the weathering of the walls of crevasses, and by separate snowfalls which have been followed by rain, hail, thawing and freezing, etc. Small rills were gliding and swirling over the melting surface with a smooth, oily appearance, in channels of pure ice — their quick, compliant movements contrasting most impressively with the rigid, invisible flow of the glacier itself, on whose back they all were riding.

Night drew near before I reached the eastern base of the mountain, and my camp lay many a rugged mile to the north; but ultimate success was assured. It was now only a matter of endurance and ordinary mountain-craft. The sunset was, if possible, yet more beautiful than that of the day before. The Mono landscape seemed to be fairly saturated with warm, purple light. The peaks marshaled along the summit were in shadow, but through every notch and pass streamed vivid sunfire, soothing and irradiating their rough, black angles, while companies of small, luminous clouds hovered above them like very angels of light.

Darkness came on, but I found my way by the trends of the cañons and the peaks projected against the sky. All excitement died with the light, and then I was weary. But the joyful sound of the waterfall across the lake was heard at last, and soon the stars were seen reflected in the lake itself. Taking my bearings from these, I discovered the little pine thicket in which my nest was, and then I had a rest such as only a tired mountaineer may enjoy. After lying loose and lost for a while, I made a sunrise fire, went down to the lake, dashed water on my head, and dipped a cupful for tea. The revival brought about by bread and tea was as complete as the exhaustion from excessive enjoyment and toil. Then I crept beneath the pine tassels to bed. The wind was frosty and the fire burned low, but my sleep was none the less sound, and the evening constellations had swept far to the west before I awoke.

After thawing and resting in the morning sunshine, I sauntered home, — that is, back to the Tuolumne camp, — bearing away toward a cluster of peaks that hold the fountain snows of one of the north tributaries of Rush Creek. Here I discovered a group of beautiful glacier lakes, nestled together in a grand amphitheater. Toward evening, I crossed the divide separating the Mono waters from those of the Tuolumne, and entered the glacier basin that now holds the fountain snows of the stream that forms the upper Tuolumne cascades. This stream I traced down through its many dells and gorges, meadows and bogs, reaching the brink of the main Tuolumne at dusk.

A loud whoop for the artists was answered again and again. Their camp-fire came in sight, and half an hour afterward I was with them. They seemed unreasonably glad to see me. I had been absent only three days; nevertheless, though the weather was fine, they had already been weighing chances as to whether I would ever return, and trying to decide whether they should wait longer or begin to seek their way back to the lowlands. Now their curious troubles were over. They packed their precious sketches, and next morning we set out homeward bound, and in two days entered the Yosemite Valley from the north by way of Indian Cañon.

THE Sierra bear, brown or gray, the sequoia of the animals, tramps over all the park, though few travelers have the pleasure of seeing him. On he fares through the majestic forests and cañons, facing all sorts of weather, rejoicing in his strength, everywhere at home, harmonizing with the trees and rocks and shaggy chaparral. Happy fellow! his lines have fallen in pleasant places, — lily gardens in silver-fir forests, miles of bushes in endless variety and exuberance of bloom over hill-waves and valleys and along the banks of streams, cañons full of music and waterfalls, parks fair as Eden, — places in which one might expect to meet angels rather than bears.

In this happy land no famine comes nigh him. All the year round his bread is sure, for some of the thousand kinds that he likes are always in season and accessible, ranged on the shelves of the mountains like stores in a pantry. From one to another, from climate to climate, up and down he climbs, feasting on each in turn, — enjoying as great variety as if he traveled to far-off countries north and

Our National Parks, 1901

south. To him almost everything is food except granite. Every tree helps to feed him, every bush and herb, with fruits and flowers, leaves and bark; and all the animals he can catch, — badgers, gophers, ground squirrels, lizards, snakes, etc., and ants, bees, wasps, old and young, together with their eggs and larvæ and nests. Craunched and hashed, down all go to his marvelous stomach, and vanish as if cast into a fire. What digestion! A sheep or a wounded deer or a pig he eats warm, about as quickly as a boy eats a buttered muffin; or should the meat be a month old, it still is welcomed with tremendous relish. After so gross a meal as this, perhaps the next will be strawberries and clover, or raspberries with mushrooms and nuts, or puckery acorns and chokecherries. And as if fearing that anything eatable in all his dominions should escape being eaten, he breaks into cabins to look after sugar, dried apples, bacon, etc. Occasionally he eats the mountaineer's bed; but when he has had a full meal of more tempting dainties he usually leaves it undisturbed, though he has been known to drag it up through a hole in the roof, carry it to the foot of a tree, and lie down on it to enjoy a siesta. Eating everything, never is he himself eaten except by man, and only man is an enemy to be feared. "B'ar meat," said a hunter from whom I was seeking information, "b'ar meat is the best meat in the mountains; their skins make the best beds, and their grease the best butter. Biscuit shortened with b'ar grease goes as far as beans; a man will walk all day on a couple of them biscuit."

In my first interview with a Sierra bear we were frightened and embarrassed, both of us, but the bear's behavior was better than mine. When I discovered him, he was standing in a narrow strip of meadow, and I was concealed behind a tree on the side of it. After studying his appearance as he stood at rest, I rushed toward him to frighten him, that I might study his gait in running. But, contrary to all I had heard about the shyness of bears, he did not run at all; and when I stopped short within a few steps of him, as he held his ground in a fighting attitude, my mistake was monstrously plain. I was then put on my good behavior, and never afterward forgot the right manners of the wilderness.

This happened on my first Sierra excursion in the forest to the north of Yosemite Valley. I was eager to meet the animals, and many of them came to me as if willing to show themselves and make my acquaintance; but the bears kept out of my way.

An old mountaineer, in reply to my questions, told me that bears were very shy, all save grim old grizzlies, and that I might travel the mountains for years without seeing one, unless I gave my mind to them and practiced the stealthy ways of hunters. Nevertheless, it was only a few weeks after I had received this information that I met the one mentioned above, and obtained instruction at first-hand.

I was encamped in the woods about a mile back of the rim of Yosemite, beside a stream that falls into the valley by the way of Indian Cañon. Nearly every day for weeks I went to the top of the North Dome to sketch; for it commands a general view of the valley, and I was anxious to draw every tree and rock and waterfall. Carlo, a St. Bernard dog, was my companion, — a fine, intelligent fellow that belonged to a hunter who was compelled to remain all summer on the hot plains, and who loaned him to me for the season for the sake of having him in the mountains, where he would be so much better off. Carlo knew bears through long experience, and he it was who led me to my first interview, though he seemed as much surprised as the bear at my unhunter-like behavior. One morning in June, just as the sunbeams began to stream through the trees, I set out for a day's sketching on the dome; and before we had gone half a mile from camp Carlo snuffed the air and looked cautiously ahead, lowered his bushy tail, drooped his ears, and began to step softly like a cat, turning every few yards and looking me in the face with a telling expression, saying plainly enough, "There is a bear a little way ahead." I walked carefully in the indicated direction, until I approached a small flowery meadow that I was familiar with, then crawled to the foot of a tree on its margin, bearing in mind what I had been told about the shyness of bears. Looking out cautiously over the instep of the tree, I saw a big, burly cinnamon bear about thirty yards off, half erect, his paws resting on the trunk of a fir that had fallen into the meadow, his hips almost buried in grass and flowers. He was listening attentively and trying to catch the scent, showing that in some way he was aware of our approach. I watched his gestures, and tried to make the most of my opportunity to learn what I could about him, fearing he would not stay long. He made a fine picture, standing alert in the sunny garden walled in by the most beautiful firs in the world.

After examining him at leisure, noting the sharp muzzle thrust inquiringly forward, the long shaggy hair on his broad chest, the stiff ears nearly buried in hair, and the slow, heavy way in which he moved his head, I foolishly made a rush on him, throwing up my arms and shouting to frighten him, to see him run. He did not mind the demonstration much; only pushed his head farther forward, and looked at me sharply as if asking, "What now? If you want to fight, I'm ready." Then I began to fear that on me would fall the work of running. But I was afraid to run, lest he should be encouraged to pursue me; therefore I held my ground, staring him in the face within a dozen yards or so, putting on as bold a look as I could, and hoping the influence of the human eye would be as great as it is said to be. Under these strained relations the interview seemed to last a long time. Finally, the bear, seeing how still I was, calmly withdrew his huge paws from the log, gave me a piercing look, as if warning me not to follow him, turned, and walked slowly up the middle of the meadow into the forest; stopping every few steps and looking back to make sure that I was not trying to take him at a disadvantage in a rear attack. I was glad to part with him, and greatly enjoyed the vanishing view as he waded through the lilies and columbines.

Thenceforth I always tried to give bears respectful notice of my approach, and they usually kept well out of my way. Though they often came around my camp in the night, only once afterward, as far as I know, was I very near one of them in daylight. This time it was a grizzly I met; and as luck would have it, I was even nearer to him than I had been to the big cinnamon. Though not a large specimen, he seemed formidable enough at a distance of less than a dozen yards. His shaggy coat was well grizzled, his head almost white. When I first caught sight of him he was eating acorns under a Kellogg oak, at a distance of perhaps seventy-five yards, and I tried to slip past without disturbing him. But he had either heard my steps on the gravel or caught my scent, for he came straight toward me, stopping every rod or so to look and listen: and as I was afraid to be seen running, I crawled on my hands and knees a little way to one side and hid behind a libocedrus, hoping he would pass me unnoticed. He soon came up opposite me, and stood looking ahead, while I looked at him, peering past the bulging trunk of the tree. At last, turning his head, he caught sight of mine, stared sharply a minute or two, and then, with fine dignity, disappeared in a manzanita-covered earthquake talus.

Considering how heavy and broad-footed bears are, it is wonderful how little harm they do in the wilderness. Even in the well-watered gardens of the middle region, where the flowers grow tallest, and where during warm weather the bears wallow and roll, no evidence of destruction is visible. On the contrary, under nature's direction, the massive beasts act as gardeners. On the forest floor, carpeted with needles and brush, and on the tough sod of glacier meadows, bears make no mark; but around the sandy margin of lakes their magnificent tracks form grand lines of embroidery. Their well-worn trails extend along the main cañons on either side, and though dusty in some places make no scar on the landscape. They bite and break off the branches of some of the pines and oaks to get the nuts, but this pruning is so light that few mountaineers ever notice it; and though they interfere with the orderly lichen-veiled decay of fallen trees, tearing them to pieces to reach the colonies of ants that inhabit them, the scattered ruins are quickly pressed back into harmony by snow and rain and over-leaning vegetation.

The number of bears that make the Park their home may be guessed by the number that have been killed by the two best hunters, Duncan and old David Brown. Duncan began to be known as a bear-killer about the year 1865. He was then roaming the woods, hunting and prospecting on the south fork of the Merced. A friend told me that he killed his first bear near his cabin at Wawona; that after mustering courage to fire he fled, without waiting to learn the effect of his shot. Going back in a few hours he found poor Bruin dead, and gained courage to try again. Duncan confessed to me, when we made an excursion together in 1875, that he was at first mortally afraid of bears, but after killing a half dozen he began to keep count of his victims, and became ambitious to be known as a great bear-hunter. In nine years he had killed forty-nine, keeping count by notches cut on one of the timbers of his cabin on the shore of Crescent Lake, near the south boundary of the Park. He said the more he knew about bears, the more he respected them and the less he feared them. But at the same time he grew more and more cautious, and never fired until he had every advantage, no matter how long he had to wait and how far he had to go before he got the bear just right as to the direction of the wind, the distance, and the way of escape in case of accident; making allowance also for the character of the animal, old or young, cinnamon or grizzly. For old grizzlies, he said, he had no use whatever, and he was mighty careful to avoid their acquaintance. He wanted to kill an even hundred; then he was going to confine himself to safer game. There was not much money in bears, anyhow, and a round hundred was enough for glory.

I have not seen or heard of him lately, and do not know how his bloody count stands. On my excursions, I occasionally passed his cabin. It was full of meat and skins hung in bundles from the rafters, and the ground about it was strewn with bones and hair, — infinitely less tidy than a bear's den. He went as hunter and guide with a geological survey party for a year or two, and was very proud of the scientific knowledge he picked up. His admiring fellow mountaineers, he said, gave him credit for knowing not only the botanical names of all the trees and bushes, but also the "botanical names of the bears."

The most famous hunter of the region was David Brown, an old pioneer, who early in the gold period established his main camp in a little forest glade on the north fork of the Merced, which is still called "Brown's Flat." No finer solitude for a hunter and prospector could be found; the climate is delightful all the year, and the scenery of both earth and sky is a perpetual feast. Though he was not much of a "scenery fellow," his friends say that he knew a pretty place when he saw it as well as any one, and liked mightily to get on the top of a commanding ridge to "look off."

When out of provisions, he would take down his old-fashioned long-barreled rifle from its deer-horn rest over the fireplace and set out in search of game. Seldom did he have to go far for venison, because the deer liked the wooded slopes of Pilot Peak ridge, with its open spots where they could rest and look about them, and enjoy the breeze from the sea in warm weather, free from troublesome flies, while they found hiding-places and fine aromatic food in the deer-brush chaparral. A small, wise dog was his only companion, and well the little mountaineer understood the object of every hunt, whether deer or bears, or only grouse hidden in the fir-tops. In deer-hunting Sandy had little to do, trotting behind his master as he walked noiselessly through the fragrant woods, careful not to step heavily on dry twigs, scanning open spots in the chaparral where the deer feed in the early morning and toward sunset, peering over ridges and swells as new outlooks were reached, and along alder and willow fringed flats and streams, until he found a young buck, killed it, tied its legs together, threw it on his shoulder, and so back to camp. But when bears were hunted, Sandy played an important part as leader, and several times saved his master's life; and it was as a bear-hunter that David Brown became famous. His method, as I had it from a friend who had passed many an evening in his cabin listening to his long stories of adventure, was simply to take a few pounds of flour and his rifle, and go slowly and silently over hill and valley in the loneliest part of the wilderness, until little Sandy came upon the fresh track of a bear, then follow it to the death, paying no heed to time. Wherever the bear went he went, how-ever rough the ground, led by Sandy, who looked back from time to time to see how his master was coming on, and regulated his face accordingly, never growing weary or allowing any other track to divert him. When high ground was reached a halt was made, to scan the openings in every direction, and perchance Bruin would be discovered sitting upright on his haunches, eating manzanita berries; pulling down the fruit-laden branches with his paws and pressing them together, so as to get sub-stantial mouthfuls, however mixed with leaves and twigs. The time of year enabled the hunter to determine approximately where the game would be found: in spring and early summer, in lush grass and clover meadows and in berry tangles along the banks of streams, or on pea-vine and lupine clad slopes; in late summer and autumn, beneath the pines, eating the cones cut off by the squirrels, and in oak groves at the bottom of cañons, munching acorns, manzanita berries, and cherries; and after snow had fallen, in alluvial bottoms, feeding on ants and yellow-jacket wasps. These food places were always cautiously approached, so as to avoid the chance of sudden encounters.

"Whenever," said the hunter, "I saw a bear before he saw me, I had no trouble in killing him. I just took lots of time to learn what he was up to and how long he would be likely to stay, and to study the direction of the wind and the lay of the land. Then I worked round to leeward of him, no matter how far I had to go; crawled and dodged to within a hundred yards, near the foot of a tree that I could climb, but which was too small for a bear to climb. There I looked well to the priming of my rifle, took off my boots so as to climb quickly if necessary, and, with my rifle in rest and Sandy behind me, waited until my bear stood right, when I made a sure, or at least a good shot back of the fore leg. In case he showed fight, I got up the tree I had in mind, before he could reach me. But bears are slow and awkward with their eyes, and being to windward they could not scent me, and often I got in a second shot before they saw the smoke. Usually, however, they tried to get away when they were hurt, and I let them go a good safe while before I ventured into the brush after them. Then Sandy was pretty sure to find them dead; if not, he barked bold as a lion to draw attention, or rushed in and nipped them behind, enabling me to get to a safe distance and watch a chance for a finishing shot.

"Oh yes, bear-hunting is a mighty interesting business, and safe enough if followed just right, though, like every other business, especially the wild kind, it has its accidents, and Sandy and I have had close calls at times. Bears are nobody's fools, and they know enough to let men alone as a general thing, unless they are wounded, or cornered, or have cubs. In my opinion, a hungry old mother would catch and eat a man, if she could; which is only fair play, anyhow, for we eat them. But nobody, as far as I know, has been eaten up in these rich mountains. Why they never tackle a fellow when he is lying asleep I never could understand. They could gobble us mighty handy, but I suppose it's nature to respect a sleeping man."

Sheep-owners and their shepherds have killed a great many bears, mostly by poison and traps of various sorts. Bears are fond of mutton, and levy heavy toll on every flock driven into the mountains. They usually come to the corral at night, climb in, kill a sheep with a stroke of the paw, carry it off a little distance, eat about half of it, and return the next night for the other half; and so on all summer, or until they are themselves killed. It is not, however, by direct killing, but by suffocation through crowding against the corral wall in fright, that the greatest losses are incurred. From ten to fifteen sheep are found dead, smothered in the corral, after every attack; or the walls are broken, and the flock is scattered far and wide. A flock may escape the attention of these marauders for a week or two in the spring; but after their first taste of the fine mountain-fed meat the visits are persistently kept up, in spite of all precautions. Once I spent a night with two Portuguese shepherds, who were greatly troubled with bears, from two to four or five visiting them almost every night. Their camp was near the middle of the Park, and the wicked bears, they said, were getting worse and worse. Not waiting now until dark, they came out of the brush in broad daylight, and boldly carried off as many sheep as they liked. One evening, before sundown, a bear, followed by two cubs, came for an early supper, as the flock was being slowly driven toward the camp. Joe, the elder of the shepherds, warned by many exciting experiences, promptly climbed a tall tamarack pine, and left the freebooters to help themselves; while Antone, calling him a coward, and declaring that he was not going to let bears eat up his sheep before his face, set the dogs on them, and rushed toward them with a great noise and a stick. The frightened cubs ran up a tree, and the mother ran to meet the shepherd and dogs. Antone stood astonished for a moment, eyeing the oncoming bear; then fled faster than Joe had, closely pursued. He scrambled to the roof of their little cabin, the only refuge quickly available; and fortunately, the bear, anxious about her young, did not climb after him, — only held him in mortal terror a few minutes, glaring and threatening, then hastened back to her cubs, called them down, went to the frightened, huddled flock, killed a sheep, and feasted in peace. Antone piteously entreated cautious Joe to show him a good safe tree, up which he climbed like a sailor climbing a mast, and held on as long as he could with legs crossed, the slim pine recommended by Joe being nearly branchless. "So you, too, are a bear coward as well as Joe," I said, after hearing the story. "Oh, I tell you," he replied, with grand solemnity, "bear face close by look awful; she just as soon eat me as not. She do so as eef all my sheeps b'long every one to her own self. I run to bear no more. I take tree every time."

After this the shepherds corraled the flock about an hour before sundown, chopped large quantities of dry wood and made a circle of fires around the corral every night, and one with a gun kept watch on a stage built in a pine by the side of the cabin, while the other slept. But after the first night or two this fire fence did no good, for the robbers seemed to regard the light as an advantage, after becoming used to it.

On the night I spent at their camp the show made by the wall of fire when it was blazing in its prime was magnificent, — the illumined trees round about relieved against solid darkness, and the two thousand sheep lying down in one gray mass, sprinkled with gloriously brilliant gems, the effect of the firelight in their eyes. It was nearly midnight when a pair of the freebooters arrived. They walked boldly through a gap in the fire circle, killed two sheep, carried them out, and vanished in the dark woods, leaving ten dead in a pile, trampled down and smothered against the corral fence; while the scared watcher in the tree did not fire a single shot, saying he was afraid he would hit some of the sheep, as the bears got among them before he could get a good sight.

In the morning I asked the shepherds why they did not move the flock to a new pasture. "Oh, no use!" cried Antone. "Look my dead sheeps. We move three four time before, all the same bear come by the track. No use. To-morrow we go home below. Look my dead sheeps. Soon all dead."

Thus were they driven out of the mountains more than a month before the usual time. After Uncle Sam's soldiers, bears are the most effective forest police, but some of the shepherds are very successful in killing them. Altogether, by hunters, mountaineers, Indians, and sheepmen, probably five or six hundred have been killed within the bounds of the Park, during the last thirty years. But they are not in danger of extinction. Now that the Park is guarded by soldiers, not only has the vegetation in great part come back to the desolate ground, but all the wild animals are increasing in numbers. No guns are allowed in the Park except under certain restrictions, and after a permit has been obtained from the officer in charge. This has stopped the barbarous slaughter of bears, and especially of deer, by shepherds, hunters, and hunting tourists, who, it would seem, can find no pleasure without blood.

The Sierra deer — the blacktail — spend the winters in the brushy and exceedingly rough region just below the main timber-belt, and are less accessible to hunters there than when they are passing through the comparatively open forests to and from their summer pastures near the summits of the range. They go up the mountains early in the spring as the snow melts, not waiting for it all to disappear; reaching the high Sierra about the first of June, and the coolest recesses at the base of the peaks a month or so later. I have tracked them for miles over compacted snow from three to ten feet deep.

Deer are capital mountaineers, making their way into the heart of the roughest mountains; seeking not only pasturage, but a cool climate, and safe hidden places in which to bring forth their young. They are not supreme as rock-climbing animals; they take second rank, yielding the first to the mountain sheep, which dwell above them on the highest crags and peaks. Still, the two meet frequently; for the deer climbs all the peaks save the lofty summits above the glaciers, crossing piles of angular boulders, roaring swollen streams, and sheer-walled cañons by fords and passes that would try the nerves of the hardiest mountaineers, — climbing with graceful ease and reserve of strength that cannot fail to arouse admiration.

152

THE ANIMALS OF THE YOSEMITE

Everywhere some species of deer seems to be at home, — on rough or smooth ground, lowlands or highlands, in swamps and barrens and the densest woods, in varying climates, hot or cold, over all the continent; maintaining glorious health, never making an awkward step. Standing, lying down, walking, feeding, running even for life, it is always invincibly graceful, and adds beauty and animation to every landscape, — a charming animal, and a great credit to nature.

I never see one of the common blacktail deer, the only species in the Park, without fresh admiration; and since I never carry a gun I see them well: lying beneath a juniper or dwarf pine, among the brown needles on the brink of some cliff or the end of a ridge commanding a wide outlook; feeding in sunny openings among chaparral, daintily selecting aromatic leaves and twigs; leading their fawns out of my way, or making them lie down and hide; bounding past through the forest, or curiously advancing and retreating again and again.

One morning when I was eating breakfast in a little garden spot on the Kaweah, hedged around with chaparral, I noticed a deer's head thrust through the bushes, the big beautiful eyes gazing at me. I kept still, and the deer ventured forward a step, then snorted and withdrew. In a few minutes she returned, and came into the open garden, stepping with infinite grace, followed by two others. After showing themselves for a moment, they bounded over the hedge with sharp, timid snorts and vanished. But curiosity brought them back with still another, and all four came into my garden, and, satisfied that I meant them no ill, began to feed, actually eating breakfast with me, like tame, gentle sheep around a shepherd, — rare company, and the most graceful in movements and attitudes. I eagerly watched them while they fed on ceanothus and wild cherry, daintily culling single leaves here and there from the side of the hedge, turning now and then to snip a few leaves of mint from the midst of the garden flowers. Grass they did not eat at all. No wonder the contents of the deer's stomach are eaten by the Indians.

While exploring the upper cañon of the north fork of the San Joaquin, one evening, the sky threatening rain, I searched for a dry bed, and made choice of a big juniper that had been pushed down by a snow avalanche, but was resting stubbornly on its knees high enough to let me lie under its broad trunk. Just below my shelter there was another juniper on the

very brink of a precipice, and, examining it, I
found a deer-bed beneath it, completely protected and concealed by drooping branches,
a fine refuge and lookout as well as resting-
place. About an hour before dark I heard the
clear, sharp snorting of a deer, and looking
down on the brushy, rocky cañon bottom, discovered an anxious doe that no doubt had her
fawns concealed near by. She bounded over
the chaparral and up the farther slope of the
wall, often stopping to look back and listen,
— a fine picture of vivid, eager alertness. I
sat perfectly still, and as my shirt was colored
like the juniper bark I was not easily seen.
After a little she came cautiously toward me,
sniffing the air and grazing, and her movements, as she descended the cañon side over
boulder piles and brush and fallen timber,
were admirably strong and beautiful; she
never strained or made apparent efforts, although jumping high here and there. As she
drew nigh she sniffed anxiously, trying the air
in different directions until she caught my
scent; then bounded off, and vanished behind
a small grove of firs. Soon she came back with
the same caution and insatiable curiosity, —
coming and going five or six times. While I
sat admiring her, a Douglas squirrel, evidently excited by her noisy alarms, climbed a

boulder beneath me, and witnessed her performances as attentively as I did, while a frisky chipmunk, too restless or hungry for such shows, busied himself about his supper in a thicket of shadbushes, the fruit of which was then ripe, glancing about on the slender twigs lightly as a sparrow.

Toward the end of the Indian summer, when the young are strong, the deer begin to gather in little bands of from six to fifteen or twenty, and on the approach of the first snowstorm they set out on their march down the mountains to their winter quarters; lingering usually on warm hillsides and spurs eight or ten miles below the summits, as if loath to leave. About the end of November, a heavy, far-reaching storm drives them down in haste along the dividing ridges between the rivers, led by old experienced bucks whose knowledge of the topography is wonderful.

It is when the deer are coming down that the Indians set out on their grand fall hunt. Too lazy to go into the recesses of the mountains away from trails, they wait for the deer to come out, and then waylay them. This plan also has the advantage of finding them in bands. Great preparations are made. Old guns are mended, bullets moulded, and the hunters wash themselves and fast to some extent, to insure good luck, as they say. Men and women, old and young, set forth together. Central camps are made on the well-known highways of the deer, which are soon red with blood. Each hunter comes in laden, old crones as well as maidens smiling on the luckiest. All grow fat and merry. Boys, each armed with an antlered head, play at buck-fighting, and plague the industrious women, who are busily preparing the meat for transportation, by stealing up behind them and throwing fresh hides over them. But the Indians are passing away here as everywhere, and their red camps on the mountains are fewer every year.

There are panthers, foxes, badgers, porcupines, and coyotes in the Park, but not in large numbers. I have seen coyotes well back in the range at the head of the Tuolumne Meadows as early as June 1st, before the snow was gone, feeding on marmots; but they are far more numerous on the inhabited lowlands around ranches, where they enjoy life on chickens, turkeys, quail eggs, ground squirrels, hares, etc., and all kinds of fruit. Few wild sheep, I fear, are left hereabouts; for, though safe on the high peaks, they are driven down the eastern slope of the mountains when the deer are driven down the western, to ridges and outlying spurs where the snow does not fall to a great depth, and there they are within reach of the cattlemen's rifles.

The two squirrels of the Park, the Douglas and the California gray, keep all the woods lively. The former is far more abundant and more widely distributed, being found all the way up from the foothills to the dwarf pines on the Summit peaks. He is the most influential of the Sierra animals, though small, and the brightest of all the squirrels I know, — a squirrel of squirrels, quick mountain vigor and valor condensed, purely wild, and as free from disease as a sunbeam. One cannot think of such an animal ever being weary or sick. He claims all the woods, and is inclined to drive away even men as intruders. How he scolds, and what faces he makes! If not so comically small he would be a dreadful fellow. The gray, *Sciurus fossor*, is the handsomest, I think, of all the large American squirrels. He is something like the Eastern gray, but is brighter and clearer in color, and more lithe and slender. He dwells in the oak and pine woods up to a height of about five thousand feet above the sea, is rather common in Yosemite Valley, Hetch-Hetchy, Kings River Cañon, and indeed in all the main cañons and Yosemites, but does not like the high fir-covered ridges. Compared with the Douglas, the gray is more than twice as large; nevertheless, he manages to make his way through the trees with less stir than his small, peppery neighbor, and is much less influential in every way. In the spring, before the pine-nuts and hazel-nuts are ripe, he examines last year's cones for the few seeds that may be left in them between the half-open scales, and gleans fallen nuts and seeds on the ground among the leaves, after making sure that no enemy is nigh. His fine tail floats, now behind, now above him, level or gracefully curled, light and radiant as dry thistledown. His body seems hardly more substantial than his tail. The Douglas is a firm, emphatic bolt of life, fiery, pungent, full of brag and show and fight, and his movements have none of the elegant deliberation of the gray. They are so quick and keen they almost sting the onlooker, and the acrobatic harlequin gyrating show he makes of himself turns one giddy to see. The gray is shy and oftentimes stealthy, as if half expecting to find an enemy in every tree and bush and behind every log; he seems to wish to be let alone, and manifests no desire to be seen, or admired, or feared. He is hunted by the Indians, and this of itself is cause enough for caution. The Douglas is less attractive for game, and probably increasing in numbers in spite of every enemy. He goes his ways bold as a lion, up and down and across, round and round, the happiest, merriest of all the hairy tribe, and at the same time tremendously earnest and solemn, sunshine incarnate, making every tree tingle with his electric toes. If you prick him, you cannot think he will bleed. He seems above the chance and change that beset common mortals, though in busily gathering burs and nuts he shows that he has to work for a living, like the rest of us. I never found a dead Douglas. He gets into the world and out of it without being noticed; only in prime is he seen, like some little plants that are visible only when in bloom.

The little striped *Tamias quadrivittatus* is one of the most amiable and delightful of all the mountain tree-climbers. A brighter, cheerier chipmunk does not exist. He is smarter, more arboreal and squirrel-like, than the familiar Eastern species, and is distributed as widely on the Sierra as the Douglas. Every forest, however dense or open, every hilltop and cañon, however brushy or bare, is cheered and enlivened by this happy little animal. You are likely to notice him first on the lower edge of the coniferous belt, where the Sabine and yellow pines meet; and thence upward, go where you may, you will find him every day, even in winter, unless the weather is stormy.

THE ANIMALS OF THE YOSEMITE

He is an exceedingly interesting little fellow, full of odd, quaint ways, confiding, thinking no evil; and without being a squirrel — a true shadow-tail — he lives the life of a squirrel, and has almost all squirrelish accomplishments without aggressive quarrelsomeness.

I never weary of watching him as he frisks about the bushes, gathering seeds and berries; poising on slender twigs of wild cherry, shad, chinquapin, buckthorn, bramble; skimming along prostrate trunks or over the grassy, needle-strewn forest floor; darting from boulder to boulder on glacial pavements and the tops of the great domes. When the seeds of the conifers are ripe, he climbs the trees and cuts off the cones for a winter store, working diligently, though not with the tremendous lightning energy of the Douglas, who frequently drives him out of the best trees. Then he lies in wait, and picks up a share of the burs cut off by his domineering cousin, and stores them beneath logs and in hollows. Few of the Sierra animals are so well liked as this little airy, fluffy half squirrel, half spermophile. So gentle, confiding, and busily cheery and happy, he takes one's heart and keeps his place among the best-loved of the mountain darlings. A diligent collector of seeds, nuts, and berries, of course he is well fed, though never in the least dumpy with fat. On the contrary, he looks like a mere fluff of fur, weighing but little more than a field mouse, and of his frisky, birdlike liveliness without haste there is no end. Douglas can bark with his mouth closed, but little quad always opens his when he talks or sings. He has a considerable variety of notes which correspond with his movements, some of them sweet and liquid, like water dripping into a pool with tinkling sound. His eyes are black and animated, shining like dew. He seems dearly to like teasing a dog, venturing within a few feet of it, then frisking away with a lively chipping and low squirrelish churring; beating time to his music, such as it is, with his tail, which at each chip and churr describes a half circle. Not even Douglas is surer footed or takes greater risks. I have seen him running about on sheer Yosemite cliffs, holding on with as little effort as a fly and as little thought of danger, in places where, if he had made the least slip, he would have fallen thousands of feet. How fine it would be could mountaineers move about on precipices with the same sure grip!

Before the pine-nuts are ripe, grass seeds and those of the many species of ceanothus, with strawberries, raspberries, and the soft red thimbleberries of *Rubus nutkanus*, form the bulk of his food, and a neater eater is not to be found in the mountains. Bees powdered with pollen, poking their blunt noses into the bells of flowers, are comparatively clumsy and boorish. Frisking along some fallen pine or fir, when the grass seeds are ripe, he looks about him, considering which of the tufts he sees is likely to have the best, runs out to it, selects what he thinks is sure to be a good head, cuts it off, carries it to the top of the log, sits upright and nibbles out the grain without getting awns in his mouth, turning the head round, holding it and fingering it as if playing on a flute; then skips for another and another, bringing them to the same dining-log.

The woodchuck (*Arctomys monax*) dwells on high bleak ridges and boulder piles; and a very different sort of mountaineer is he, — bulky, fat, aldermanic, and fairly bloated at times by hearty indulgence in the lush pastures of his airy home. And yet he is by no means a dull animal. In the midst of what we regard as storm-beaten desolation, high in the frosty air, beside the glaciers he pipes and whistles right cheerily and lives to a good old age. If you are as early a riser as he is, you may oftentimes⋅ see him come blinking out of his burrow to meet the first beams of the morning and take a sunbath on some favorite flat-topped boulder. Afterward, well warmed, he goes to breakfast in one of his garden hollows, eats heartily like a cow in clover until comfortably swollen, then goes a-visiting, and plays and loves and fights.

In the spring of 1875, when I was exploring the peaks and glaciers about the head of the middle fork of the San Joaquin, I had crossed the range from the head of Owen River, and one morning, passing around a frozen lake where the snow was perhaps ten feet deep, I was surprised to find the fresh track of a woodchuck plainly marked, the sun having softened the surface. What could the animal be thinking of, coming out so early while all the ground was snow-buried? The steady trend of his track showed he had a definite aim, and fortunately it was toward a mountain thirteen thousand feet high that I meant to climb. So I followed to see if I could find out what he was up to. From the base of the mountain the track pointed straight up, and I knew by the melting snow that I was not far behind him. I lost the track on a crumbling ridge, partly projecting through the snow, but soon discovered it again. Well toward the summit of the mountain, in an open spot on the south side, nearly inclosed by disintegrating pinnacles among which the sun heat reverberated, making an isolated patch of warm climate, I found a nice garden, full of rock cress, phlox, silene, draba, etc., and a few grasses; and in this garden I overtook the wanderer, enjoying a fine fresh meal, perhaps the first of the season. How did he know the way to this one garden spot, so high and far off, and what told him that it was in bloom while yet the snow was ten feet deep over his den? For this it would seem he would need more botanical, topographical, and climatological knowledge than most mountaineers are possessed of.

The shy, curious mountain beaver, *Haplodon*, lives on the heights, not far from the woodchuck. He digs canals and controls the flow of small streams under the sod. And it is startling when one is camped on the edge of a sloping meadow near the homes of these industrious mountaineers, to be awakened in the still night by the sound of water rushing and gurgling under one's head in a newly formed canal. Pouched gophers also have a way of awakening nervous campers that is quite as exciting as the haplodon's plan; that is, by a series of firm upward pushes when they are driving tunnels and shoving up the dirt. One naturally cries out, "Who's there?" and then discovering the cause, "All right. Go on. Good-night," and goes to sleep again.

The haymaking pika, bob-tailed spermophile, and wood-rat are also among the most interesting of the Sierra animals. The last, *Neotoma*, is scarcely at all like the common rat, is nearly twice as large, has a delicate, soft, brownish fur, white on the belly, large ears thin and translucent, eyes full and liquid and mild in expression, most blunt and squirrelish, slender claws sharp as needles, and as his limbs are strong he can climb about as well as a squirrel; while no rat or squirrel has so innocent a look, is so easily approached, or in general expresses so much confidence in one's good intentions. He seems too fine for the thorny thickets he inhabits, and his big, rough hut is as unlike himself as possible. No other animal in these mountains makes nests so large and striking in appearance as his. They are built of all kinds of sticks (broken branches, and old rotten moss-grown chunks and green twigs, smooth or thorny, cut from the nearest bushes), mixed with miscellaneous rubbish and curious odds and ends, — bits of cloddy earth, stones, bones, bits of deer-horn, etc.: the whole simply piled in conical masses on the ground in chaparral thickets. Some of these cabins are five or six feet high, and occasionally a dozen or more are grouped together; less, perhaps, for society's sake than for advantages of food and shelter.

THE ANIMALS OF THE YOSEMITE

Coming through deep, stiff chaparral in the heart of the wilderness, heated and weary in forcing a way, the solitary explorer, happening into one of these curious neotoma villages, is startled at the strange sight, and may imagine he is in an Indian village, and feel anxious as to the reception he will get in a place so wild. At first, perhaps, not a single inhabitant will be seen, or at most only two or three seated on the tops of their huts as at the doors, observing the stranger with the mildest of mild eyes. The nest in the center of the cabin is made of grass and films of bark chewed to tow, and lined with feathers and the down of various seeds. The thick, rough walls seem to be built for defense against enemies — fox, coyote, etc. — as well as for shelter, and the delicate creatures in their big, rude homes, suggest tender flowers, like those of *Salvia carduacea*, defended by thorny involucres.

Sometimes the home is built in the forks of an oak, twenty or thirty feet from the ground, and even in garrets. Among housekeepers who have these bushmen as neighbors or guests they are regarded as thieves, because they carry away and pile together everything transportable (knives, forks, tin cups, spoons, spectacles, combs, nails, kindling-wood, etc., as well as eatables of all sorts), to strengthen their fortifications or to shine among rivals. Once, far back in the high Sierra, they stole my snow-goggles, the lid of my teapot, and my aneroid barometer; and one stormy night, when encamped under a prostrate cedar, I was awakened by a gritting sound on the granite, and by the light of my fire I discovered a handsome neotoma beside me, dragging away my ice-hatchet, pulling with might and main by a buckskin string on the handle. I threw bits of bark at him and made a noise to frighten him, but he stood scolding and chattering back at me, his fine eyes shining with an air of injured innocence.

A great variety of lizards enliven the warm portions of the Park. Some of them are more than a foot in length, others but little larger than grasshoppers. A few are snaky and repulsive at first sight, but most of the species are handsome and attractive, and bear acquaintance well; we like them better the farther we see into their charming lives. Small fellow mortals, gentle and guileless, they are easily tamed, and have beautiful eyes, expressing the clearest innocence, so that, in spite of prejudices brought from cool, lizardless countries, one must soon learn to like them. Even the horned toad of the plains and foothills, called horrid, is mild and gentle, with charming eyes, and so are the snakelike species found in the underbrush of the lower forests. These glide in curves with all the ease and grace of snakes, while their small, undeveloped limbs drag for the most part as useless appendages. One specimen that I measured was fourteen inches long, and as far as I saw it made no use whatever of its diminutive limbs.

Most of them glint and dart on the sunny rocks and across open spaces from bush to bush, swift as dragonflies and hummingbirds, and about as brilliantly colored. They never make a long-sustained run, whatever their object, but dart direct as arrows for a distance of ten or twenty feet, then suddenly stop, and as suddenly start again. These stops are necessary as rests, for they are short-winded, and when pursued steadily are soon run out of breath, pant pitifully, and may easily be caught where no retreat in bush or rock is quickly available.

If you stay with them a week or two and behave well, these gentle saurians, descendants of an ancient race of giants, will soon know and trust you, come to your feet, play, and watch your every motion with cunning curiosity. You will surely learn to like them, not only the bright ones, gorgeous as the rainbow, but the little ones, gray as lichened granite, and scarcely bigger than grasshoppers; and they will teach you that scales may cover as fine a nature as hair or feathers or anything tailored.

There are many snakes in the cañons and lower forests, but they are mostly handsome and harmless. Of all the tourists and travelers who have visited Yosemite and the adjacent mountains, not one has been bitten by a snake of any sort, while thousands have been charmed by them. Some of them vie with the lizards in beauty of color and dress patterns. Only the rattlesnake is venomous, and he carefully keeps his venom to himself as far as man is concerned, unless his life is threatened.

Before I learned to respect rattlesnakes I killed two, the first on the San Joaquin plain. He was coiled comfortably around a tuft of bunch-grass, and I discovered him when he was between my feet as I was stepping over him. He held his head down and did not attempt to strike, although in danger of being trampled. At that time, thirty years ago, I imagined that rattlesnakes should be killed wherever found. I had no weapon of any sort, and on the smooth plain there was not a stick or a stone within miles; so I crushed him by jumping on him, as the deer are said to do. Looking me in the face he saw I meant mischief, and quickly cast himself into a coil, ready to strike in defense. I knew he could not strike when traveling, therefore I threw handfuls of dirt and grass sods at him, to tease him out of coil. He held his ground a few minutes, threatening and striking, and then started off to get rid of me. I ran forward and jumped on him; but he drew back his head so quickly my heel missed, and he also missed his stroke at me. Persecuted, tormented, again and again he tried to get away, bravely striking out to protect himself; but at last my heel came squarely down, sorely wounding him, and a few more brutal stampings crushed him. I felt degraded by the killing business, farther from heaven, and I made up my mind to try to be at least as fair and charitable as the snakes themselves, and to kill no more save in self-defense.

The second killing might also, I think, have been avoided, and I have always felt somewhat sore and guilty about it. I had built a little cabin in Yosemite, and for convenience in getting water, and for the sake of music and society, I led a small stream from Yosemite Creek into it. Running along the side of the wall it was not in the way, and it had just fall enough to ripple and sing in low, sweet tones, making delightful company, especially at night when I was lying awake. Then a few frogs came in and made merry with the stream, — and one snake, I suppose to catch the frogs.

Returning from my long walks, I usually brought home a large handful of plants, partly for study, partly for ornament, and set them in a corner of the cabin, with their stems in the stream to keep them fresh. One day, when I picked up a handful that had begun to fade, I uncovered a large coiled rattler that had been hiding behind the flowers. Thus suddenly brought to light face to face with the rightful owner of the place, the poor reptile was desperately embarrassed, evidently realizing that he had no right in the cabin. It was not only fear that he showed, but a good deal of downright bashfulness and embarrassment, like that of a more than half honest person caught under suspicious circumstances behind a door. Instead of striking or threatening to strike, though coiled and ready, he slowly drew his head down as far as he could, with awkward, confused kinks in his neck and a shamefaced expression, as if wishing the ground would open and hide him. I have looked into the eyes of so many wild animals that I feel sure I did not mistake the feelings of this unfortunate snake. I did not want to kill him, but I had many visitors, some of them children, and I oftentimes came in late at night; so I judged he must die.

THE ANIMALS OF THE YOSEMITE

Since then I have seen perhaps a hundred or more in these mountains, but I have never intentionally disturbed them, nor have they disturbed me to any great extent, even by accident, though in danger of being stepped on. Once, while I was on my knees kindling a fire, one glided under the arch made by my arm. He was only going away from the ground I had selected for a camp, and there was not the slightest danger, because I kept still and allowed him to go in peace. The only time I felt myself in serious danger was when I was coming out of the Tuolumne Cañon by a steep side cañon toward the head of Yosemite Creek. On an earthquake talus, a boulder in my way presented a front so high that I could just reach the upper edge of it while standing on the next below it. Drawing myself up, as soon as my head was above the flat top of it I caught sight of a coiled rattler. My hands had alarmed him, and he was ready for me; but even with this provocation, and when my head came in sight within a foot of him, he did not strike. The last time I sauntered through the big cañon I saw about two a day. One was not coiled, but neatly folded in a narrow space between two cobblestones on the side of the river, his head below the level of them, ready to shoot up like a jack-in-the-box for frogs or birds. My foot spanned the space above within an inch or two of his head, but he only held it lower. In making my way through a particularly tedious tangle of buckthorn, I parted the branches on the side of an open spot and threw my bundle of bread into it; and when, with my arms free, I was pushing through after it, I saw a small rattlesnake dragging his tail from beneath my bundle. When he caught sight of me he eyed me angrily, and with an air of righteous indignation seemed to be asking why I had thrown that stuff on him. He was so small that I was inclined to slight him, but he struck out so angrily that I drew back, and approached the opening from the other side. But he had been listening, and when I looked through the brush I found him confronting me, still with a come-in-if-you-dare expression. In vain I tried to explain that I only wanted my bread; he stoutly held the ground in front of it; so I went back a dozen rods and kept still for half an hour, and when I returned he had gone.

One evening, near sundown, in a very rough, boulder-choked portion of the cañon, I searched long for a level spot for a bed, and at last was glad to find a patch of flood-sand on the river-bank, and a lot of driftwood close by for a camp-fire. But when I threw down my bundle,

I found two snakes in possession of the ground. I might have passed the night even in this snake den without danger, for I never knew a single instance of their coming into camp in the night; but fearing that, in so small a space, some late comers, not aware of my presence, might get stepped on when I was replenishing the fire, to avoid possible crowding I encamped on one of the earthquake boulders.

There are two species of *Crotalus* in the Park, and when I was exploring the basin of Yosemite Creek I thought I had discovered a new one. I saw a snake with curious divided appendages on its head. Going nearer, I found that the strange headgear was only the feet of a frog. Cutting a switch, I struck the snake lightly until he disgorged the poor frog, or rather allowed it to back out. On its return to the light from one of the very darkest of death valleys, it blinked a moment with a sort of dazed look, then plunged into a stream, apparently happy and well.

Frogs abound in all the bogs, marshes, pools, and lakes, however cold and high and isolated. How did they manage to get up these high mountains? Surely not by jumping. Long and dry excursions through weary miles of boulders and brush would be trying to frogs. Most likely their stringy spawn is carried on the feet of ducks, cranes, and other water-birds. Anyhow, they are most thoroughly distributed, and flourish famously. What a cheery, hearty set they are, and how bravely their krink and tronk concerts enliven the rocky wilderness!

None of the high-lying mountain lakes or branches of the rivers above sheer falls had fish of any sort until stocked by the agency of man. In the high Sierra, the only river in which trout exist naturally is the middle fork of Kings River. There are no sheer falls on this stream; some of the rapids, however, are so swift and rough, even at the lowest stage of water, that it is surprising any fish can climb them. I found trout in abundance in this fork up to seventy-five hundred feet. They also run quite high on the Kern. On the Merced they get no higher than Yosemite Valley, four thousand feet, all the forks of the river being barred there by sheer falls, and on the main Tuolumne they are stopped by a fall below Hetch-Hetchy, still lower than Yosemite. Though these upper waters are inaccessible to the fish, one would suppose their eggs might have been planted there by some means. Nature has so many ways of doing such things. In this case she waited for the agency of man, and now many of these hitherto fishless lakes and streams are full of fine trout, stocked by individual enterprise, Walton clubs, etc., in great part under the auspices of the United States Fish Commission. A few trout carried into Hetch-Hetchy in a common water-bucket have multiplied wonderfully fast. Lake Tenaya, at an elevation of over eight thousand feet, was stocked eight years ago by Mr. Murphy, who carried a few trout from Yosemite. Many of the small streams of the eastern slope have also been stocked with trout transported over the passes in tin cans on the backs of mules. Soon, it would seem, all the streams of the range will be enriched by these lively fish, and will become the means of drawing thousands of visitors into the mountains. Catching trout with a bit of bent wire is a rather trivial business, but fortunately people fish better than they know. In most cases it is the man who is caught. Trout-fishing regarded as bait for catching men, for the saving of both body and soul, is important, and deserves all the expense and care bestowed on it.

THE YELLOWSTONE NATIONAL PARK

Of the four national parks of the West, the Yellowstone is far the largest. It is a big, wholesome wilderness on the broad summit of the Rocky Mountains, favored with abundance of rain and snow, — a place of fountains where the greatest of the American rivers take their rise. The central portion is a densely forested and comparatively level volcanic plateau with an average elevation of about eight thousand feet above the sea, surrounded by an imposing host of mountains belonging to the subordinate Gallatin, Wind River, Teton, Absaroka, and snowy ranges. Unnumbered lakes shine in it, united by a famous band of streams that rush up out of hot lava beds, or fall from the frosty peaks in channels rocky and bare, mossy and bosky, to the main rivers, singing cheerily on through every difficulty, cunningly dividing and finding their way east and west to the two far-off seas.

Glacier meadows and beaver meadows are outspread with charming effect along the banks of the streams, parklike expanses in the woods, and innumerable small gardens in

Our National Parks, 1901

rocky recesses of the mountains, some of them containing more petals than leaves, while the whole wilderness is enlivened with happy animals.

Beside the treasures common to most mountain regions that are wild and blessed with a kind climate, the park is full of exciting wonders. The wildest geysers in the world, in bright, triumphant bands, are dancing and singing in it amid thousands of boiling springs, beautiful and awful, their basins arrayed in gorgeous colors like gigantic flowers; and hot paint-pots, mud springs, mud volcanoes, mush and broth caldrons whose contents are of every color and consistency, plash and heave and roar in bewildering abundance. In the adjacent mountains, beneath the living trees the edges of petrified forests are exposed to view, like specimens on the shelves of a museum, standing on ledges tier above tier where they grew, solemnly silent in rigid crystalline beauty after swaying in the winds thousands of centuries ago, opening marvelous views back into the years and climates and life of the past. Here, too, are hills of sparkling crystals, hills of sulphur, hills of glass, hills of cinders and ashes, mountains of every style of architecture, icy or forested, mountains covered with honey-bloom sweet as Hymettus, mountains boiled soft like potatoes and colored like a sunset sky. A' that and a' that, and twice as muckle 's a' that, Nature has on show in the Yellowstone Park. Therefore it is called Wonderland, and thousands of tourists and travelers stream into it every summer, and wander about in it enchanted.

Fortunately, almost as soon as it was discovered it was dedicated and set apart for the benefit of the people, a piece of legislation that shines benignly amid the common dust-and-ashes history of the public domain, for which the world must thank Professor Hayden above all others; for he led the first scientific exploring party into it, described it, and with admirable enthusiasm urged Congress to preserve it. As delineated in the year 1872, the park contained about 3344 square miles. On March 30, 1891, it was to all intents and purposes enlarged by the Yellowstone National Park Timber Reserve, and in December, 1897, by the Teton Forest Reserve; thus nearly doubling its original area, and extending the southern boundary far enough to take in the sublime Teton range and the famous pasture lands of the big Rocky Mountain game animals. The withdrawal of this large tract from the public domain did no harm to any one; for its height, six thousand to over thirteen thousand feet above the sea, and its thick mantle of volcanic rocks, prevent its ever being available for agriculture or mining, while on the other hand its geographical position, reviving climate, and wonderful scenery combine to make it a grand health, pleasure, and study resort, — a gathering-place for travelers from all the world.

The national parks are not only withdrawn from sale and entry like the forest reservations, but are efficiently managed and guarded by small troops of United States cavalry, directed by the Secretary of the Interior. Under this care the forests are flourishing, protected from both axe and fire; and so, of course, are the shaggy beds of underbrush and the herbaceous vegetation. The so-called curiosities, also, are preserved, and the furred and feathered tribes, many of which, in danger of extinction a short time ago, are now increasing in numbers, — a refreshing thing to see amid the blind, ruthless destruction that is going on in the adjacent regions. In pleasing contrast to the noisy, ever-changing management, or mismanagement, of blundering, plundering, money-making vote-sellers who receive their places from boss politicians as purchased goods, the soldiers do their duty so quietly that the traveler is scarce aware of their presence.

This is the coolest and highest of the parks.
Frosts occur every month of the year. Nevertheless, the tenderest tourist finds it warm
enough in summer. The air is electric and full
of ozone, healing, reviving, exhilarating, kept
pure by frost and fire, while the scenery is wild
enough to awaken the dead. It is a glorious
place to grow in and rest in; camping on the
shores of the lakes, in the warm openings of
the woods golden with sunflowers, on the
banks of the streams, by the snowy waterfalls,
beside the exciting wonders or away from them
in the scallops of the mountain walls sheltered
from every wind, on smooth silky lawns enameled with gentians, up in the fountain hollows
of the ancient glaciers between the peaks,
where cool pools and brooks and gardens of
precious plants charmingly embowered are
never wanting, and good rough rocks with
every variety of cliff and scaur are invitingly
near for outlooks and exercise.

From these lovely dens you may make excursions whenever you like into the middle of the
park, where the geysers and hot springs are
reeking and spouting in their beautiful basins,
displaying an exuberance of color and strange
motion and energy admirably calculated to
surprise and frighten, charm and shake up the
least sensitive out of apathy into newness of life.

YELLOWSTONE NATIONAL PARK

However orderly your excursions or aimless, again and again amid the calmest, stillest scenery you will be brought to a standstill hushed and awe-stricken before phenomena wholly new to you. Boiling springs and huge deep pools of purest green and azure water, thousands of them, are plashing and heaving in these high, cool mountains as if a fierce furnace fire were burning beneath each one of them; and a hundred geysers, white torrents of boiling water and steam, like inverted waterfalls, are ever and anon rushing up out of the hot, black underworld. Some of these ponderous geyser columns are as large as sequoias, — five to sixty feet in diameter, one hundred and fifty to three hundred feet high, — and are sustained at this great height with tremendous energy for a few minutes, or perhaps nearly an hour, standing rigid and erect, hissing, throbbing, booming, as if thunderstorms were raging beneath their roots, their sides roughened or fluted like the furrowed boles of trees, their tops dissolving in feathery branches, while the irised spray, like misty bloom is at times blown aside, revealing the massive shafts shining against a background of pine-covered hills. Some of them lean more or less, as if storm-bent, and instead of being round are flat or fan-shaped, issuing from irregular slits in silex pavements with radiate structure, the sunbeams sifting through them in ravishing splendor. Some are broad and round-headed like oaks; others are low and bunchy, branching near the ground like bushes; and a few are hollow in the centre like big daisies or water-lilies. No frost cools them, snow never covers them nor lodges in their branches; winter and summer they welcome alike; all of them, of whatever form or size, faithfully rising and sinking in fairy rhythmic dance night and day, in all sorts of weather, at varying periods of minutes, hours, or weeks, growing up rapidly, uncontrollable as fate, tossing their pearly branches in the wind, bursting into bloom and vanishing like the frailest flowers, — plants of which Nature raises hundreds or thousands of crops a year with no apparent exhaustion of the fiery soil.

The so-called geyser basins, in which this rare sort of vegetation is growing, are mostly open valleys on the central plateau that were eroded by glaciers after the greater volcanic fires had ceased to burn. Looking down over the forests as you approach them from the surrounding heights, you see a multitude of white columns, broad, reeking masses, and irregular jets and puffs of misty vapor ascending from the bottom of the valley, or entangled like smoke among the neighboring trees, suggesting the factories of some busy town or the camp-fires of an army. These mark the position of each mush-pot, paint-pot, hot spring, and geyser, or gusher, as the Icelandic words mean. And when you saunter into the midst of them over the bright sinter pavements, and see how pure and white and pearly gray they are in the shade of the mountains, and how radiant in the sunshine, you are fairly enchanted. So numerous they are and varied, Nature seems to have gathered them from all the world as specimens of her rarest fountains, to show in one place what she can do. Over four thousand hot springs have been counted in the park, and a hundred geysers; how many more there are nobody knows.

These valleys at the heads of the great rivers may be regarded as laboratories and kitchens, in which, amid a thousand retorts and pots, we may see Nature at work as chemist or cook, cunningly compounding an infinite variety of mineral messes; cooking whole mountains; boiling and steaming flinty rocks to smooth paste and mush, — yellow, brown, red, pink, lavender, gray, and creamy white, — making the most beautiful mud in the world; and distilling the most ethereal essences. Many of these pots and caldrons have been boiling thousands of years. Pots of sulphurous mush, stringy and lumpy, and pots of broth as black as ink, are tossed and stirred with constant care, and thin transparent essences, too pure and fine to be called water, are kept simmering gently in beautiful sinter cups and bowls that grow ever more beautiful the longer they are used. In some of the spring basins, the waters, though still warm, are perfectly calm, and shine blandly in a sod of overleaning grass and flowers, as if they were thoroughly cooked at last, and set aside to settle and cool. Others are wildly boiling over as if running to waste, thousands of tons of the precious liquids being thrown into the air to fall in scalding floods on the clean coral floor of the establishment, keeping onlookers at a distance. Instead of holding limpid pale green or azure water, other pots and craters are filled with scalding mud, which is tossed up from three or four feet to thirty feet, in sticky, rank-smelling masses, with gasping, belching, thudding sounds, plastering the branches of neighboring trees; every flask, retort, hot spring, and geyser has something special in it, no two being the same in temperature, color, or composition.

In these natural laboratories one needs stout faith to feel at ease. The ground sounds hollow underfoot, and the awful subterranean thunder shakes one's mind as the ground is shaken, especially at night in the pale moonlight, or when the sky is overcast with stormclouds. In the solemn gloom, the geysers, dimly visible, look like monstrous dancing ghosts, and their wild songs and the earthquake thunder replying to the storms overhead seem doubly terrible, as if divine government were at an end. But the trembling hills keep their places. The sky clears, the rosy dawn is reassuring, and up comes the sun like a god, pouring his faithful beams across the mountains and forest, lighting each peak and tree and ghastly geyser alike, and shining into the eyes of the reeking springs, clothing them with rainbow light, and dissolving the seeming chaos of darkness into varied forms of harmony. The ordinary work of the world goes on. Gladly we see the flies dancing in the sunbeams, birds feeding their young, squirrels gathering nuts, and hear the blessed ouzel singing confidingly in the shallows of the river, — most faithful evangel, calming every fear, reducing everything to love.

The variously tinted sinter and travertine formations, outspread like pavements over large areas of the geyser valleys, lining the spring basins and throats of the craters, and forming beautiful coral-like rims and curbs about them, always excite admiring attention; so also does the play of the waters from which they are deposited. The various minerals in them are rich in colors, and these are greatly heightened by a smooth, silky growth of brilliantly colored confervæ which lines many of the pools and channels and terraces. No bed of flower-bloom is more exquisite than these myriads of minute plants, visible only in mass, growing in the hot waters. Most of the spring borders are low and daintily scalloped, crenelated, and beaded with sinter pearls. Some of the geyser craters are massive and picturesque, like ruined castles or old burned-out sequoia stumps, and are adorned on a grand scale with outbulging, cauliflower-like formations. From these as centers the silex pavements slope gently away in thin, crusty, overlapping layers, slightly interrupted in some places by low terraces. Or, as in the case of the Mammoth Hot Springs, at the north end of the park, where the building waters issue from the side of a steep hill, the deposits form a succession of higher and broader terraces of white travertine tinged with purple, like the famous Pink Terrace at Rotomahana, New Zealand, draped in front with clustering stalactites, each terrace having a pool of indescribably beautiful water upon it in a basin with a raised rim that glistens with confervæ, — the whole, when viewed at a distance of a mile or two, looking like a broad, massive cascade pouring over shelving rocks in snowy purpled foam.

The stones of this divine masonry, invisible particles of lime or silex, mined in quarries no eye has seen, go to their appointed places in gentle, tinkling, transparent currents or through the dashing turmoil of floods, as surely guided as the sap of plants streaming into bole and branch, leaf and flower. And thus from century to century this beauty-work has gone on and is going on.

Passing through many a mile of pine and spruce woods, toward the center of the park you come to the famous Yellowstone Lake. It is about twenty miles long and fifteen wide, and lies at a height of nearly eight thousand feet above the level of the sea, amid dense black forests and snowy mountains. Around its winding, wavering shores, closely forested and picturesquely varied with promontories and bays, the distance is more than one hundred miles. It is not very deep, only from two hundred to three hundred feet, and contains less water than the celebrated Lake Tahoe of the California Sierra, which is nearly the same size, lies at a height of sixty-four hundred feet, and is over sixteen hundred feet deep. But no other lake in North America of equal area lies so high as the Yellowstone, or gives birth to so noble a river. The terraces around its shores show that at the close of the glacial period its surface was about one hundred and sixty feet higher than it is now, and its area nearly twice as great.

It is full of trout, and a vast multitude of birds — swans, pelicans, geese, ducks, cranes, herons, curlews, plovers, snipe — feed in it and upon its shores; and many forest animals come out of the woods, and wade a little way in shallow, sandy places to drink and look about them, and cool themselves in the free flowing breezes.

In calm weather it is a magnificent mirror for the woods and mountains and sky, now pattered with hail and rain, now roughened with sudden storms that send waves to fringe the shores and wash its border of gravel and sand. The Absaroka Mountains and the Wind River Plateau on the east and south pour their gathered waters into it, and the river issues from the north side in a broad, smooth, stately current, silently gliding with such serene majesty that one fancies it knows the vast journey of four thousand miles that lies before it, and the work it has to do. For the first twenty miles its course is in a level, sunny valley lightly fringed with trees, through which it flows in silvery reaches stirred into spangles here and there by ducks and leaping trout, making no sound save a low whispering among the pebbles and the dipping willows and sedges of its banks. Then suddenly, as if preparing for hard work, it rushes eagerly, impetuously forward rejoicing in its strength, breaks into foam-bloom, and goes thundering down into the Grand Cañon in two magnificent falls, one hundred and three hundred feet high.

The cañon is so tremendously wild and impressive that even these great falls cannot hold your attention. It is about twenty miles long and a thousand feet deep, — a weird, unearthly-looking gorge of jagged, fantastic architecture, and most brilliantly colored. Here the Washburn range, forming the northern rim of the Yellowstone basin, made up mostly of beds of rhyolite decomposed by the action of thermal waters, has been cut through and laid open to view by the river; and a famous section it has made. It is not the depth or the shape of the cañon, nor the waterfall, nor the green and gray river chanting its brave song as it goes foaming on its way, that most impresses the observer, but the colors of the decomposed volcanic rocks. With few exceptions, the traveler in strange lands finds that, however much the scenery and vegetation in different countries may change, Mother Earth is ever familiar and the same. But here the very ground is changed, as if belonging to some other world. The walls of the cañon from top to bottom burn in a perfect glory of color, confounding and dazzling when the sun is shining, — white, yellow, green, blue, vermilion, and various other shades of red indefinitely blending. All the earth hereabouts seems to be paint. Millions of tons of it lie in sight, exposed to wind and weather as if of no account, yet marvelously fresh and bright, fast colors not to be washed out or bleached out by either sunshine or storms. The effect is so novel and awful, we imagine that even a river might be afraid to enter such a place. But the rich and gentle beauty of the vegetation is reassuring. The lovely *Linnæa borealis* hangs her twin bells over the brink of the cliffs, forests and gardens extend their treasures in smiling confidence on either side, nuts and berries ripen well whatever may be going on below; blind fears vanish, and the grand gorge seems a kindly, beautiful part of the general harmony, full of peace and joy and good will.

The park is easy of access. Locomotives

drag you to its northern boundary at Cinnabar, and horses and guides do the rest. From Cinnabar you will be whirled in coaches along the foaming Gardiner River to Mammoth Hot Springs; thence through woods and meadows, gulches and ravines along branches of the Upper Gallatin, Madison, and Firehole rivers to the main geyser basins; thence over the Continental Divide and back again, up and down through dense pine, spruce, and fir woods to the magnificent Yellowstone Lake, along its northern shore to the outlet, down the river to the falls and Grand Cañon, and thence back through the woods to Mammoth Hot Springs and Cinnabar; stopping here and there at the so-called points of interest among the geysers, springs, paint-pots, mud volcanoes, etc., where you will be allowed a few minutes or hours to saunter over the sinter pavements, watch the play of a few of the geysers, and peer into some of the most beautiful and terrible of the craters and pools. These wonders you will enjoy, and also the views of the mountains, especially the Gallatin and Absaroka ranges, the long, willowy glacier and beaver meadows, the beds of violets, gentians, phloxes, asters, phacelias, goldenrods, eriogonums, and many other flowers, some species giving color to whole meadows and hillsides. And you will enjoy your short views of the great lake and river and cañon. No scalping Indians will you see. The Blackfeet and Bannocks that once roamed here are gone; so are the old beaver-catchers, the Coulters and Bridgers, with all their attractive buckskin and romance. There are several bands of buffaloes in the park, but you will not thus cheaply in tourist fashion see them nor many of the other large animals hidden in the wilderness. The song-birds, too, keep mostly out of sight of the rushing tourist, though off the roads thrushes, warblers, orioles, grosbeaks, etc., keep the air sweet and merry. Perhaps in passing rapids and falls you may catch glimpses of the water-ouzel, but in the whirling noise you will not hear his song. Fortunately, no road noise frightens the Douglas squirrel, and his merry play and gossip will amuse you all through the woods. Here and there a deer may be seen crossing the road, or a bear. Most likely, however, the only bears you will see are the half tame ones that go to the hotels every night for dinner-table scraps, — yeast-powder biscuit, Chicago canned stuff, mixed pickles, and beefsteaks that have proved too tough for the tourists.

Among the gains of a coach trip are the acquaintances made and the fresh views into human nature; for the wilderness is a shrewd touchstone, even thus lightly approached, and brings many a curious trait to view. Setting out, the driver cracks his whip, and the four horses go off at half gallop, half trot, in trained, showy style, until out of sight of the hotel. The coach is crowded, old and young side by side, blooming and fading, full of hope and fun and care. Some look at the scenery or the horses, and all ask questions, an odd mixed lot of them: "Where is the umbrella? What is the name of that blue flower over there? Are you sure the little bag is aboard? Is that hollow yonder a crater? How is your throat this morning? How high did you say the geysers spout? How does the elevation affect your head? Is that a geyser reeking over there in the rocks, or only a hot spring?" A long ascent is made, the solemn mountains come to view, small cares are quenched, and all become natural and silent, save perhaps some unfortunate expounder who has been reading guidebook geology, and rumbles forth foggy subsidences and upheavals until he is in danger of being heaved overboard. The driver will give you the names of the peaks and meadows and streams as you come to them, call attention to the glass road, tell how hard it was to build, — how the obsidian cliffs naturally pushed the surveyor's lines to the right, and the industrious beavers, by flooding the valley in front of the cliff, pushed them to the left.

Geysers, however, are the main objects, and as soon as they come in sight other wonders are forgotten. All gather around the crater of the one that is expected to play first. During the eruptions of the smaller geysers, such as the Beehive and Old Faithful, though a little frightened at first, all welcome the glorious show with enthusiasm, and shout, "Oh, how wonderful, beautiful, splendid, majestic!" Some venture near enough to stroke the column with a stick, as if it were a stone pillar or a tree, so firm and substantial and permanent it seems. While tourists wait around a large geyser, such as the Castle or the Giant, there is a chatter of small talk in anything but solemn mood; and during the intervals between the preliminary splashes and upheavals some adventurer occasionally looks down the throat of the crater, admiring the silex formations and wondering whether Hades is as beautiful. But when, with awful uproar as if avalanches were falling and storms thundering in the depths, the tremendous outburst begins, all run away to a safe distance, and look on, awe-stricken and silent, in devout, worshiping wonder.

The largest and one of the most wonderfully

beautiful of the springs is the Prismatic, which the guide will be sure to show you. With a circumference of three hundred yards, it is more like a lake than a spring. The water is pure deep blue in the center, fading to green on the edges, and its basin and the slightly terraced pavement about it are astonishingly bright and varied in color. This one of the multitude of Yellowstone fountains is of itself object enough for a trip across the continent. No wonder that so many fine myths have originated in springs; that so many fountains were held sacred in the youth of the world, and had miraculous virtues ascribed to them. Even in these cold, doubting, questioning, scientific times many of the Yellowstone fountains seem able to work miracles. Near the Prismatic Spring is the great Excelsior Geyser, which is said to throw a column of boiling water sixty to seventy feet in diameter to a height of from fifty to three hundred feet, at irregular periods. This is the greatest of all the geysers yet discovered anywhere. The Firehole River, which sweeps past it, is, at ordinary stages, a stream about one hundred yards wide and three feet deep; but when the geyser is in eruption, so great is the quantity of water discharged that the volume of the river is doubled, and it is rendered too hot and rapid to be forded.

Geysers are found in many other volcanic regions, — in Iceland, New Zealand, Japan, the Himalayas, the Eastern Archipelago, South America, the Azores, and elsewhere; but only in Iceland, New Zealand, and this Rocky Mountain park do they display their grandest forms, and of these three famous regions the Yellowstone is easily first, both in the number and in the size of its geysers. The greatest height of the column of the Great Geyser of Iceland actually measured was two hundred and twelve feet, and of the Strokhr one hundred and sixty-two feet.

In New Zealand, the Te Pueia at Lake Taupo, the Waikite at Rotorna, and two others are said to lift their waters occasionally to a height of one hundred feet, while the celebrated Te Tarata at Rotomahana sometimes lifts a boiling column twenty feet in diameter to a height of sixty feet. But all these are far surpassed by the Excelsior. Few tourists, however, will see the Excelsior in action, or a thousand other interesting features of the park that lie beyond the wagon-roads and the hotels. The regular trips — from three to five days — are too short. Nothing can be done well at a speed of forty miles a day. The multitude of mixed, novel impressions rapidly piled on one another make only a dreamy, bewildering, swirling blur, most of which is unrememberable. Far more time should be taken. Walk away quietly in any direction and taste the freedom of the mountaineer. Camp out among the grass and gentians of glacier meadows, in craggy garden nooks full of Nature's darlings. Climb the mountains and get their good tidings. Nature's peace will flow into you as sunshine flows into trees. The winds will blow their own freshness into you, and the storms their energy, while cares will drop off like autumn leaves. As age comes on, one source of enjoyment after another is closed, but Nature's sources never fail. Like a generous host, she offers here brimming cups in endless variety, served in a grand hall, the sky its ceiling, the mountains its walls, decorated with glorious paintings and enlivened with bands of music ever playing. The petty discomforts that beset the awkward guest, the unskilled camper, are quickly forgotten, while all that is precious remains. Fears vanish as soon as one is fairly free in the wilderness.

Most of the dangers that haunt the unseasoned citizen are imaginary; the real ones are perhaps too few rather than too many for his good. The bears that always seem to spring up thick as trees, in fighting, devouring attitudes before the frightened tourist whenever a camping trip is proposed, are gentle now, finding they are no longer likely to be shot; and rattlesnakes, the other big irrational dread of over-civilized people, are scarce here, for most of the park lies above the snake-line. Poor creatures, loved only by their Maker, they are timid and bashful, as mountaineers know; and though perhaps not possessed of much of that charity that suffers long and is kind, seldom, either by mistake or by mishaps, do harm to any one. Certainly they cause not the hundredth part of the pain and death that follow the footsteps of the admired Rocky Mountain trapper. Nevertheless, again and again, in season and out of season, the question comes up, "What are rattlesnakes good for?" As if nothing that does not obviously make for the benefit of man had any right to exist; as if our ways were God's ways. Long ago, an Indian to whom a French traveler put this old question replied that their tails were good for toothache, and their heads for fever. Anyhow, they are all, head and tail, good for themselves, and we need not begrudge them their share of life.

Fear nothing. No town park you have been accustomed to saunter in is so free from danger as the Yellowstone. It is a hard place to leave.

YELLOWSTONE NATIONAL PARK

Even its names in your guidebook are attractive, and should draw you far from wagon-roads, — all save the early ones, derived from the infernal regions: Hell Roaring River, Hell Broth Springs, The Devil's Caldron, etc. Indeed, the whole region was at first called Coulter's Hell, from the fiery brimstone stories told by trapper Coulter, who left the Lewis and Clark expedition and wandered through the park, in the year 1807, with a band of Bannock Indians. The later names, many of which we owe to Mr. Arnold Hague of the U.S. Geological Survey, are so telling and exhilarating that they set our pulses dancing and make us begin to enjoy the pleasures of excursions ere they are commenced. Three River Peak, Two Ocean Pass, Continental Divide, are capital geographical descriptions, suggesting thousands of miles of rejoicing streams and all that belongs to them. Big Horn Pass, Bison Peak, Big Game Ridge, bring brave mountain animals to mind. Birch Hills, Garnet Hills, Amethyst Mountain, Storm Peak, Electric Peak, Roaring Mountain, are bright, bracing names. Wapiti, Beaver, Tern, and Swan lakes conjure up fine pictures, and so also do Osprey and Ouzel falls. Antelope Creek, Otter, Mink, and Grayling creeks, Geode, Jasper, Opal, Carnelian, and

Chalcedony creeks, are lively and sparkling names that help the streams to shine; and Azalea, Stellaria, Arnica, Aster, and Phlox creeks, what pictures these bring up! Violet, Morning Mist, Hygeia, Beryl, Vermilion, and Indigo springs, and many beside, give us visions of fountains more beautifully arrayed than Solomon in all his purple and golden glory. All these and a host of others call you to camp. You may be a little cold some nights on mountain tops above the timber-line, but you will see the stars, and by and by you can sleep enough in your town bed, or at least in your grave. Keep awake while you may in mountain mansions so rare.

If you are not very strong, try to climb Electric Peak when a big bossy, well-charged thunder-cloud is on it, to breathe the ozone set free, and get yourself kindly shaken and shocked. You are sure to be lost in wonder and praise, and every hair of your head will stand up and hum and sing like an enthusiastic congregation.

After this reviving experience, you should take a look into a few of the tertiary volumes of the grand geological library of the park, and see how God writes history. No technical knowledge is required; only a calm day and a calm mind. Perhaps nowhere else in the

YELLOWSTONE NATIONAL PARK

Rocky Mountains have the volcanic forces been so busy. More than ten thousand square miles hereabouts have been covered to a depth of at least five thousand feet with material spouted from chasms and craters during the tertiary period, forming broad sheets of basalt, andesite, rhyolite, etc., and marvelous masses of ashes, sand, cinders, and stones now consolidated into conglomerates, charged with the remains of plants and animals that lived in the calm, genial periods that separated the volcanic outbursts.

Perhaps the most interesting and telling of these rocks, to the hasty tourist, are those that make up the mass of Amethyst Mountain. On its north side it presents a section two thousand feet high of roughly stratified beds of sand, ashes, and conglomerates coarse and fine, forming the untrimmed edges of a wonderful set of volumes lying on their sides, — books a million years old, well bound, miles in size, with full-page illustrations. On the ledges of this one section we see trunks and stumps of fifteen or twenty ancient forests ranged one above another, standing where they grew, or prostrate and broken like the pillars of ruined temples in desert sands, — a forest fifteen or twenty stories high, the roots of each spread above the tops of the next beneath it, telling wonderful tales of the bygone centuries, with their winters and summers, growth and death, fire, ice, and flood.

There were giants in those days. The largest of the standing opal and agate stumps and prostrate sections of the trunks are from two or three to fifty feet in height or length, and from five to ten feet in diameter; and so perfect is the petrifaction that the annual rings and ducts are clearer and more easily counted than those of living trees, centuries of burial having brightened the records instead of blurring them. They show that the winters of the tertiary period gave as decided a check to vegetable growth as do those of the present time. Some trees favorably located grew rapidly, increasing twenty inches in diameter in as many years, while others of the same species, on poorer soil or overshadowed, increased only two or three inches in the same time.

Among the roots and stumps on the old forest floors we find the remains of ferns and bushes, and the seeds and leaves of trees like those now growing on the southern Alleghanies, — such as magnolia, sassafras, laurel, linden, persimmon, ash, alder, dogwood. Studying the lowest of these forests, the soil it grew on and the deposits it is buried in, we see that it was rich in species, and flourished in a genial, sunny climate. When its stately trees were in their glory, volcanic fires broke forth from chasms and craters, like larger geysers, spouting ashes, cinders, stones, and mud, which fell on the doomed forest like hail and snow; sifting, hurtling through the leaves and branches, choking the streams, covering the ground, crushing bushes and ferns, rapidly deepening, packing around the trees and breaking them, rising higher until the topmost boughs of the giants were buried, leaving not a leaf or twig in sight, so complete was the desolation. At last the volcanic storm began to abate, the fiery soil settled; mud floods and boulder floods passed over it, enriching it, cooling it; rains fell and mellow sunshine, and it became fertile and ready for another crop. Birds, and the winds, and roaming animals brought seeds from more fortunate woods, and a new forest grew up on the top of the buried one. Centuries of genial growing seasons passed. The seedling trees became giants, and with strong outreaching branches spread a leafy canopy over the gray land.

The sleeping subterranean fires again awake and shake the mountains, and every leaf trembles. The old craters, with perhaps new ones, are opened, and immense quantities of ashes, pumice, and cinders are again thrown into the sky. The sun, shorn of his beams, glows like a dull red ball, until hidden in sulphurous clouds. Volcanic snow, hail, and floods fall on the new forest, burying it alive, like the one beneath its roots. Then come another noisy band of mud floods and boulder floods, mixing, settling, enriching the new ground, more seeds, quickening sunshine and showers; and a third noble magnolia forest is carefully raised on the top of the second. And so on. Forest was planted above forest and destroyed, as if Nature were ever repenting, undoing the work she had so industriously done, and burying it.

Of course this destruction was creation, progress in the march of beauty through death. How quickly these old monuments excite and hold the imagination! We see the old stone stumps budding and blossoming and waving in the wind as magnificent trees, standing shoulder to shoulder, branches interlacing in grand varied round-headed forests; see the sunshine of morning and evening gilding their mossy trunks, and at high noon spangling on the thick glossy leaves of the magnolia, filtering through translucent canopies of linden and ash, and falling in mellow patches on the ferny floor; see the shining after rain, breathe the exhaling fragrance, and hear the winds and birds and the murmur of brooks and insects.

We watch them from season to season; see the swelling buds when the sap begins to flow in the spring, the opening leaves and blossoms, the ripening of summer fruits, the colors of autumn, and the maze of leafless branches and sprays in winter; and we see the sudden oncome of the storms that overwhelmed them.

One calm morning at sunrise I saw the oaks and pines in Yosemite Valley shaken by an earthquake, their tops swishing back and forth, and every branch and needle shuddering as if in distress like the frightened screaming birds. One many imagine the trembling, rocking, tumultuous waving of those ancient Yellowstone woods, and the terror of their inhabitants when the first foreboding shocks were felt, the sky grew dark, and rock-laden floods began to roar. But though they were close pressed and buried, cut off from sun and wind, all their happy leaf-fluttering and waving done, other currents coursed through them, fondling and thrilling every fibre, and beautiful wood was replaced by beautiful stone. Now their rocky sepulchres are partly open, and show forth the natural beauty of death.

After the forest times and fire times had passed away, and the volcanic furnaces were banked and held in abeyance, another great change occurred. The glacial winter came on. The sky was again darkened, not with dust and ashes, but with snow which fell in glorious abundance, piling deeper, deeper, slipping from the overladen heights in booming avalanches, compacting into glaciers, that flowed over all the landscape, wiping off forests, grinding, sculpturing, fashioning the comparatively featureless lava beds into the beautiful rhythm of hill and dale and ranges of mountains we behold to-day; forming basins for lakes, channels for streams, new soils for forests, gardens, and meadows. While this ice-work was going on, the slumbering volcanic fires were boiling the subterranean waters, and with curious chemistry decomposing the rocks, making beauty in the darkness; these forces, seemingly antagonistic, working harmoniously together. How wild their meetings on the surface were we may imagine. When the glacier period began, geysers and hot springs were playing in grander volume, it may be, than those of to-day. The glaciers flowed over them while they spouted and thundered, carrying away their fine sinter and travertine structures, and shortening their mysterious channels.

The soils made in the down-grinding required to bring the present features of the landscape into relief are possibly no better than were some of the old volcanic soils that were carried away, and which, as we have seen, nourished magnificent forests, but the glacial landscapes are incomparably more beautiful than the old volcanic ones were. The glacial winter has passed away, like the ancient summers and fire periods, though in the chronology of the geologist all these times are recent. Only small residual glaciers on the cool northern slopes of the highest mountains are left of the vast all-embracing ice-mantle, as solfataras and geysers are all that are left of the ancient volcanoes.

Now the post-glacial agents are at work on the grand old palimpsest of the park region, inscribing new characters; but still in its main telling features it remains distinctly glacial. The moraine soils are being leveled, sorted, refined, re-formed, and covered with vegetation; the polished pavements and scoring and other superficial glacial inscriptions on the crumbling lavas are being rapidly obliterated; gorges are being cut in the decomposed rhyolites and loose conglomerates, and turrets and pinnacles seem to be springing up like growing trees; while the geysers are depositing miles of sinter and travertine. Nevertheless, the ice-work is scarce blurred as yet. These later effects are only spots and wrinkles on the grand glacial countenance of the park.

Perhaps you have already said that you have seen enough for a lifetime. But before you go away you should spend at least one day and a night on a mountain top, for a last general, calming, settling view. Mount Washburn is a good one for the purpose, because it stands in the middle of the park, is unencumbered with other peaks, and is so easy of access that the climb to its summit is only a saunter. First your eye goes roving around the mountain rim amid the hundreds of peaks: some with plain flowing skirts, others abruptly precipitous and defended by sheer battlemented escarpments; flat-topped or round; heaving like sea-waves or spired and turreted like Gothic cathedrals; streaked with snow in the ravines, and darkened with files of adventurous trees climbing the ridges. The nearer peaks are perchance clad in sapphire blue, others far off in creamy white. In the broad glare of noon they seem to shrink and crouch to less than half their real stature, and grow dull and uncommunicative, — mere dead, draggled heaps of waste ashes and stone, giving no hint of the multitude of animals enjoying life in their fastnesses, or of the bright bloom-bordered streams and lakes. But when storms blow they awake and arise, wearing robes of cloud and mist in majestic speaking attitudes like gods. In the color glory of morning and evening they become still more impressive; steeped in the divine light of the alpenglow their earthiness disappears, and, blending with the heavens, they seem neither high nor low.

Over all the central plateau, which from here seems level, and over the foothills and lower slopes of the mountains, the forest extends like a black uniform bed of weeds, interrupted only by lakes and meadows and small burned spots called parks, — all of them, except the Yellowstone Lake, being mere dots and spangles in general views, made conspicuous by their color and brightness. About eighty-five per cent of the entire area of the park is covered with trees, mostly the indomitable lodge-pole pine (*Pinus contorta*, var. *Murrayana*), with a few patches and sprinklings of Douglas spruce, Engelmann spruce, silver fir (*Abies lasiocarpa*), *Pinus flexilis*, and a few alders, aspens, and birches. The Douglas spruce is found only on the lowest portions, the silver fir on the highest, and the Engelmann spruce on the dampest places, best defended from fire. Some fine specimens of the flexilis pine are growing on the margins of openings, — wide-branching, sturdy trees, as broad as high, with trunks five feet in diameter, leafy and shady, laden with purple cones and rose-colored flowers. The Engelmann spruce and sub-alpine silver fir are beautiful and notable trees, — tall, spiry, hardy, frost and snow defying, and widely distributed over the West, wherever there is a mountain to climb or a cold moraine slope to cover. But neither of these is a good fire-fighter. With rather thin bark, and scattering their seeds every year as soon as they are ripe, they are quickly driven out of fire-swept regions. When the glaciers were melting, these hardy mountaineering trees were probably among the first to arrive on the new moraine soil beds; but as the plateau became drier and fires began to run, they were driven up the mountains, and into the wet spots and islands where we now find them, leaving nearly all the park to the lodge-pole pine, which though as thin-skinned as they and as easily killed by fire, takes pains to store up its seeds in firmly closed cones, and holds them from three to nine years, so that, let the fire come when it may, it is ready to die and ready to live again in a new generation. For when the killing fires have devoured the leaves and thin resinous bark, many of the cones, only scorched, open as soon as the smoke clears away; the hoarded store of seeds is sown broadcast on the cleared ground, and a new growth immediately springs up triumphant out of the ashes. Therefore, this tree not only holds its ground, but extends its conquests farther after every fire. Thus the evenness and closeness of its growth are accounted for. In one part of the forest that I examined, the growth was about as close as a cane-brake. The trees were from four to eight inches in diameter, one hundred feet high, and one hundred and seventy-five years old. The lower limbs die young and drop off for want of light. Life with these close-planted trees is a race for light, more light, and so they push straight for the sky. Mowing off ten feet from the top of the forest would make it look like a crowded mass of telegraph-poles; for only the sunny tops are leafy. A sapling ten years old, growing in the sunshine, has as many leaves as a crowded tree one or two hundred years old. As fires are multiplied and the mountains become drier, this wonderful lodge-pole pine bids fair to obtain possession of nearly all the forest ground in the West.

How still the woods seem from here, yet how lively a stir the hidden animals are making; digging, gnawing, biting, eyes shining, at work and play, getting food, rearing young, roving through the underbrush, climbing the rocks, wading solitary marshes, tracing the banks of the lakes and streams! Insect swarms are dancing in the sunbeams, burrowing in the ground, diving, swimming, — a cloud of witnesses telling Nature's joy. The plants are as busy as the animals, every cell in a swirl of enjoyment, humming like a hive, singing the old new song of creation. A few columns and puffs of steam are seen rising above the tree-tops, some near, but most of them far off, indicating geysers and hot springs, gentle-looking and noiseless as downy clouds, softly hinting the reaction going on between the surface and the hot interior. From here you see them better than when you are standing beside them, frightened and confused, regarding them as lawless cataclysms. The shocks and outbursts of earthquakes, volcanoes, geysers, storms, the pounding of waves, the uprush of sap in plants, each and all tell the orderly love-beats of Nature's heart.

Turning to the eastward, you have the Grand Cañon and reaches of the river in full view; and yonder to the southward lies the great lake, the largest and most important of all the high fountains of the Missouri-Mississippi, and the last to be discovered.

In the year 1541, when De Soto, with a romantic band of adventurers, was seeking gold and glory and the fountain of youth, he found the Mississippi a few hundred miles above its mouth, and made his grave beneath its floods. La Salle, in 1682, after discovering the Ohio, one of the largest and most beautiful branches of the Mississippi, traced the latter to the sea from the mouth of the Illinois, through adventures and privations not easily realized now. About the same time Joliet and Father Marquette reached the "Father of Waters" by way of the Wisconsin, but more than a century passed ere its highest sources in these mountains were seen. The advancing stream of civilization has ever followed its guidance toward the west, but none of the thousand tribes of Indians living on its banks could tell the explorer whence it came. From those romantic De Soto and La Salle days to these times of locomotives and tourists, how much has the great river seen and done! Great as it now is, and still growing longer through the ground of its delta and the basins of receding glaciers at its head, it was immensely broader toward the close of the glacial period, when the ice-mantle of the mountains was melting: then with its three hundred thousand miles of branches outspread over the plains and valleys of the continent, laden with fertile mud, it made the biggest and most generous bed of soil in the world.

WILDERNESS ESSAYS

Think of this mighty stream springing in the first place in vapor from the sea, flying on the wind, alighting on the mountains in hail and snow and rain, lingering in many a fountain feeding the trees and grass; then gathering its scattered waters, gliding from its noble lake, and going back home to the sea, singing all the way! On it sweeps, through the gates of the mountains, across the vast prairies and plains, through many a wild, gloomy forest, cane-brake, and sunny savanna; from glaciers and snowbanks and pine woods to warm groves of magnolia and palm; geysers dancing at its head keeping time with the sea-waves at its mouth; roaring and gray in rapids, booming in broad, bossy falls, murmuring, gleaming in long, silvery reaches, swaying now hither, now thither, whirling, bending in huge doubling, eddying folds, serene, majestic, ungovernable, overflowing all its metes and bounds, frightening the dwellers upon its banks; building, wasting, uprooting, planting; engulfing old islands and making new ones, taking away fields and towns as if in sport, carrying canoes and ships of commerce in the midst of its spoils and drift, fertilizing the continent as one vast farm. Then, its work done, it gladly vanishes in its ocean home, welcomed by the waiting waves.

YELLOWSTONE NATIONAL PARK

Thus naturally, standing here in the midst of its fountains, we trace the fortunes of the great river. And how much more comes to mind as we overlook this wonderful wilderness! Fountains of the Columbia and Colorado lie before us, interlaced with those of the Yellowstone and Missouri, and fine it would be to go with them to the Pacific; but the sun is already in the west, and soon our day will be done.

Yonder is Amethyst Mountain, and other mountains hardly less rich in old forests, which now seem to spring up again in their glory; and you see the storms that buried them — the ashes and torrents laden with boulders and mud, the centuries of sunshine, and the dark, lurid nights. You see again the vast floods of lava, red-hot and white-hot, pouring out from gigantic geysers, usurping the basins of lakes and streams, absorbing or driving away their hissing, screaming waters, flowing around hills and ridges, submerging every subordinate feature. Then you see the snow and glaciers taking possession of the land, making new landscapes. How admirable it is that, after passing through so many vicissitudes of frost and fire and flood, the physiognomy and even the complexion of the landscape should still be so divinely fine!

Thus reviewing the eventful past, we see Nature working with enthusiasm like a man, blowing her volcanic forges like a blacksmith blowing his smithy fires, shoving glaciers over the landscapes like a carpenter shoving his planes, clearing, ploughing, harrowing, irrigating, planting, and sowing broadcast like a farmer and gardener, doing rough work and fine work, planting sequoias and pines, rose-bushes and daisies; working in gems, filling every crack and hollow with them; distilling fine essences; painting plants and shells, clouds, mountains, all the earth and heavens, like an artist, — ever working toward beauty higher and higher. Where may the mind find more stimulating, quickening pasturage? A thousand Yellowstone wonders are calling, "Look up and down and round about you!" And a multitude of still, small voices may be heard directing you to look through all this transient, shifting show of things called "substantial" into the truly substantial spiritual world whose forms flesh and wood, rock and water, air and sunshine, only veil and conceal, and to learn that here is heaven and the dwelling-place of the angels.

The sun is setting; long, violet shadows are growing out over the woods from the mountains along the western rim of the park; the

Absaroka range is baptized in the divine light of the alpenglow, and its rocks and trees are transfigured. Next to the light of the dawn on high mountain tops, the alpenglow is the most impressive of all the terrestrial manifestations of God.

Now comes the gloaming. The alpenglow is fading into earthy, murky gloom, but do not let your town habits draw you away to the hotel. Stay on this good fire-mountain and spend the night among the stars. Watch their glorious bloom until the dawn, and get one more baptism of light. Then, with fresh heart, go down to your work, and whatever your fate, under whatever ignorance or knowledge you may afterward chance to suffer, you will remember these fine, wild views, and look back with joy to your wanderings in the blessed old Yellowstone Wonderland.

A GREAT STORM IN UTAH [1]

UTAH has just been blessed with one of the grandest storms I have ever beheld this side of the Sierra. The mountains are laden with fresh snow; wild streams are swelling and booming adown the cañons, and out in the valley of the Jordan a thousand rain-pools are gleaming in the sun.

With reference to the development of fertile storms bearing snow and rain, the greater portion of the calendar springtime of Utah has been winter. In all the upper cañons of the mountains the snow is now from five to ten feet deep or more, and most of it has fallen since March. Almost every other day during the last three weeks small local storms have been falling on the Wahsatch and Oquirrh Mountains, while the Jordan Valley remained dry and sun-filled. But on the afternoon of Thursday, the 17th ultimo, wind, rain, and snow filled the whole basin, driving wildly over valley and plain from range to range, bestowing their benefactions in most cordial and

[1] Letter dated "Salt Lake City, Utah, May 19, 1877."
San Francisco Bulletin, (May, 1877)

harmonious storm-measures. The oldest Saints
say they have never witnessed a more violent
storm of this kind since the first settlement of
Zion, and while the gale from the northwest,
with which the storm began, was rocking their
adobe walls, uprooting trees and darkening the
streets with billows of dust and sand, some of
them seemed inclined to guess that the terrible
phenomenon was one of the signs of the times
of which their preachers are so constantly re-
minding them, the beginning of the outpouring
of the treasured wrath of the Lord upon the
Gentiles for the killing of Joseph Smith. To
me it seemed a cordial outpouring of Nature's
love; but it is easy to differ with salt Latter-
Days in everything — storms, wives, politics,
and religion.

About an hour before the storm reached the
city I was so fortunate as to be out with a
friend on the banks of the Jordan enjoying the
scenery. Clouds, with peculiarly restless and
self-conscious gestures, were marshaling themselves along the mountain-tops, and sending
out long, overlapping wings across the valley;
and even where no cloud was visible, an ob-
scuring film absorbed the sunlight, giving rise
to a cold, bluish darkness. Nevertheless, distant objects along the boundaries of the landscape were revealed with wonderful distinct-

ness in this weird, subdued, cloud-sifted light. The mountains, in particular, with the forests on their flanks, their mazy lacelike cañons, the wombs of the ancient glaciers, and their marvelous profusion of ornate sculpture, were most impressively manifest. One would fancy that a man might be clearly seen walking on the snow at a distance of twenty or thirty miles.

While we were reveling in this rare, ungarish grandeur, turning from range to range, studying the darkening sky and listening to the still small voices of the flowers at our feet, some of the denser clouds came down, crowning and wreathing the highest peaks and dropping long gray fringes whose smooth linear structure showed that snow was beginning to fall. Of these partial storms there were soon ten or twelve, arranged in two rows, while the main Jordan Valley between them lay as yet in profound calm. At 4.30 P.M. a dark brownish cloud appeared close down on the plain towards the lake, extending from the northern extremity of the Oquirrh Range in a northeasterly direction as far as the eye could reach. Its peculiar color and structure excited our attention without enabling us to decide certainly as to its character, but we were not left long in doubt, for in a few minutes it came sweeping over the valley in wild uproar, a torrent of wind thick with sand and dust, advancing with a most majestic front, rolling and overcombing like a gigantic sea-wave. Scarcely was it in plain sight ere it was upon us, racing across the Jordan, over the city, and up the slopes of the Wahsatch, eclipsing all the landscapes in its course — the bending trees, the dust streamers, and the wild onrush of everything movable giving it an appreciable visibility that rendered it grand and inspiring.

This gale portion of the storm lasted over an hour, then down came the blessed rain and the snow all through the night and the next day, the snow and rain alternating and blending in the valley. It is long since I have seen snow coming into a city. The crystal flakes falling in the foul streets was a pitiful sight.

Notwithstanding the vaunted refining influences of towns, purity of all kinds — pure hearts, pure streams, pure snow — must here be exposed to terrible trials. City Creek, coming from its high glacial fountains, enters the streets of this Mormon Zion pure as an angel, but how does it leave it? Even roses and lilies in gardens most loved are tainted with a thousand impurities as soon as they unfold. I heard Brigham Young in the Tabernacle the other day warning his people that if they did not mend their manners angels would not come into their houses, though perchance they might be sauntering by with little else to do than chat with them. Possibly there may be Salt Lake families sufficiently pure for angel society, but I was not pleased with the reception they gave the small snow angels that God sent among them the other night. Only the children hailed them with delight. The old Latter-Days seemed to shun them. I should like to see how Mr. Young, the Lake Prophet, would meet such messengers.

But to return to the storm. Toward the evening of the 18th it began to wither. The snowy skirts of the Wahsatch Mountains appeared beneath the lifting fringes of the clouds, and the sun shone out through colored windows, producing one of the most glorious after-storm effects I ever witnessed. Looking across the Jordan, the gray sagey slopes from the base of the Oquirrh Mountains were covered with a thick, plushy cloth of gold, soft and ethereal as a cloud, not merely tinted and gilded like a rock with autumn sunshine, but deeply muffled beyond recognition. Surely nothing in heaven, nor any mansion of the Lord in all his worlds, could be more gloriously carpeted. Other portions of the plain were flushed with red and purple, and all the mountains and the clouds above them were painted in corresponding

loveliness. Earth and sky, round and round
the entire landscape, was one ravishing reve-
lation of color, infinitely varied and inter-
blended.

I have seen many a glorious sunset beneath
lifting storm-clouds on the mountains, but
nothing comparable with this. I felt as if new-
arrived in some other far-off world. The mountains, the plains, the sky, all seemed new.
Other experiences seemed but to have prepared
me for this, as souls are prepared for heaven.
To describe the colors on a single mountain
would, if it were possible at all, require many
a volume — purples, and yellows, and deli-
cious pearly grays divinely toned and inter-
blended, and so richly put on one seemed to be
looking down through the ground as through a
sky. The disbanding clouds lingered lovingly
about the mountains, filling the cañons like
tinted wool, rising and drooping around the
topmost peaks, fondling their rugged bases, or,
sailing alongside, trailed their lustrous fringes
through the pines as if taking a last view of
their accomplished work. Then came darkness, and the glorious day was done.

This afternoon the Utah mountains and valleys seem to belong to our own very world
again. They are covered with common sunshine. Down here on the banks of the Jordan,

larks and redwings are swinging on the rushes; the balmy air is instinct with immortal life; the wild flowers, the grass, and the farmers' grain are fresh as if, like the snow, they had come out of heaven, and the last of the angel clouds are fleeing from the mountains.

WILD WOOL

MORAL improvers have calls to preach. I have a friend who has a call to plough, and woe to the daisy sod or azalea thicket that falls under the savage redemption of his keen steel shares. Not content with the so-called subjugation of every terrestrial bog, rock, and moorland, he would fain discover some method of reclamation applicable to the ocean and the sky, that in due calendar time they might be brought to bud and blossom as the rose. Our efforts are of no avail when we seek to turn his attention to wild roses, or to the fact that both ocean and sky are already about as rosy as possible — the one with stars, the other with dulse, and foam, and wild light. The practical developments of his culture are orchards and clover-fields wearing a smiling, benevolent aspect, truly excellent in their way, though a near view discloses something barbarous in them all. Wildness charms not my friend, charm it never so wisely: and whatsoever may be the character of his heaven, his earth seems only a chaos of agricultural possibilities calling for grubbing-hoes and manures.

Sometimes I venture to approach him with a plea for wildness, when he good-naturedly shakes a big mellow apple in my face, reiterating his favorite aphorism, "Culture is an orchard apple; Nature is a crab." Not all culture, however, is equally destructive and inappreciative. Azure skies and crystal waters find loving recognition, and few there be who would welcome the axe among mountain pines, or would care to apply any correction to the tones and costumes of mountain waterfalls. Nevertheless, the barbarous notion is almost universally entertained by civilized man, that there is in all the manufactures of Nature something essentially coarse which can and must be eradicated by human culture. I was, therefore, delighted in finding that the wild wool growing upon mountain sheep in the neighborhood of Mount Shasta was much finer than the average grades of cultivated wool. This *fine* discovery was made some three months ago,[1] while hunting among the Shasta sheep between Shasta and Lower Klamath Lake. Three fleeces were obtained — one that belonged to a large ram about four years old, another to a ewe about the same age, and another to a

[1] This essay was written early in 1875. [Editor.]

yearling lamb. After parting their beautiful wool on the side and many places along the back, shoulders, and hips, and examining it closely with my lens, I shouted: "Well done for wildness! Wild wool is finer than tame!"

My companions stooped down and examined the fleeces for themselves, pulling out tufts and ringlets, spinning them between their fingers, and measuring the length of the staple, each in turn paying tribute to wildness. It *was* finer, and no mistake; finer than Spanish Merino. Wild wool *is* finer than tame.

"Here," said I, "is an argument for fine wildness that needs no explanation. Not that such arguments are by any means rare, for all wildness is finer than tameness, but because fine wool is appreciable by everybody alike — from the most speculative president of national wool-growers' associations all the way down to the gude-wife spinning by her ingleside."

Nature is a good mother, and sees well to the clothing of her many bairns — birds with smoothly imbricated feathers, beetles with shining jackets, and bears with shaggy furs. In the tropical south, where the sun warms like a fire, they are allowed to go thinly clad; but in the snowy northland she takes care to clothe warmly. The squirrel has socks and mittens, and a tail broad enough for a blanket; the grouse is densely feathered down to the ends of his toes; and the wild sheep, besides his undergarment of fine wool, has a thick overcoat of hair that sheds off both the snow and the rain. Other provisions and adaptations in the dresses of animals, relating less to climate than to the more mechanical circumstances of life, are made with the same consummate skill that characterizes all the love-work of Nature. Land, water, and air, jagged rocks, muddy ground, sand-beds, forests, underbrush, grassy plains, etc., are considered in all their possible combinations while the clothing of her beautiful wildlings is preparing. No matter what the circumstances of their lives may be, she never allows them to go dirty or ragged. The mole, living always in the dark and in the dirt, is yet as clean as the otter or the wave-washed seal; and our wild sheep, wading in snow, roaming through bushes, and leaping among jagged storm-beaten cliffs, wears a dress so exquisitely adapted to its mountain life that it is always found as unruffled and stainless as a bird.

On leaving the Shasta hunting-grounds I selected a few specimen tufts, and brought them away with a view to making more leisurely examinations; but, owing to the imperfectness of the instruments at my command, the results thus far obtained must be regarded only as rough approximations.

As already stated, the clothing of our wild sheep is composed of fine wool and coarse hair. The hairs are from about two to four inches long, mostly of a dull bluish-gray color, though varying somewhat with the seasons. In general characteristics they are closely related to the hairs of the deer and antelope, being light, spongy, and elastic, with a highly polished surface, and though somewhat ridged and spiraled, like wool, they do not manifest the slightest tendency to felt or become taggy. A hair two and a half inches long, which is perhaps near the average length, will stretch about one fourth of an inch before breaking. The diameter decreases rapidly both at the top and bottom, but is maintained throughout the greater portion of the length with a fair degree of regularity. The slender tapering point in which the hairs terminate is nearly black: but, owing to its fineness as compared with the main trunk, the quantity of blackness is not sufficient to affect greatly the general color. The number of hairs growing upon a square inch is about ten thousand; the number of wool fibers is about twenty-five thousand, or two and a half times that of the hairs. The wool fibers are white and glossy, and beautifully spired into ringlets. The average length of the staple is about an inch and a half. A fiber of this length, when growing undisturbed down among the hairs, measures about an inch; hence the degree of curliness may easily be inferred. I regret exceedingly that my instruments do not enable me to measure the diameter of the fibers, in order that their degrees of fineness might be definitely compared with each other and with the finest of the domestic breeds; but that the three wild fleeces under consideration are considerably finer than the average grades of Merino shipped from San Francisco is, I think, unquestionable.

When the fleece is parted and looked into with a good lens, the skin appears of a beautiful pale-yellow color, and the delicate wool fibers are seen growing up among the strong hairs, like grass among stalks of corn, every individual fiber being protected about as specially and effectively as if inclosed in a separate husk. Wild wool is too fine to stand by itself, the fibers being about as frail and invisible as the floating threads of spiders, while the hairs against which they lean stand erect like hazel wands; but, notwithstanding their great dissimilarity in size and appearance, the wool

and hair are forms of the same thing, modified
in just that way and to just that degree that
renders them most perfectly subservient to the
well-being of the sheep. Furthermore, it will
be observed that these wild modifications are
entirely distinct from those which are brought
chancingly into existence through the accidents and caprices of culture; the former being
inventions of God for the attainment of definite ends. Like the modifications of limbs —
the fin for swimming, the wing for flying, the
foot for walking — so the fine wool for warmth,
the hair for additional warmth and to protect
the wool, and both together for a fabric to
wear well in mountain roughness and wash
well in mountain storms.

The effects of human culture upon wild wool
are analogous to those produced upon wild
roses. In the one case there is an abnormal
development of petals at the expense of the
stamens, in the other an abnormal development of wool at the expense of the hair.
Garden roses frequently exhibit stamens in
which the transmutation to petals may be
observed in various stages of accomplishment,
and analogously the fleeces of tame sheep
occasionally contain a few wild hairs that are
undergoing transmutation to wool. Even wild
wool presents here and there a fiber that

appears to be in a state of change. In the course of my examinations of the wild fleeces mentioned above, three fibers were found that were wool at one end and hair at the other. This, however, does not necessarily imply imperfection, or any process of change similar to that caused by human culture. Water-lilies contain parts variously developed into stamens at one end, petals at the other, as the constant and normal condition. These half wool, half hair fibers may therefore subserve some fixed requirement essential to the perfection of the whole, or they may simply be the fine boundary-lines where an exact balance between the wool and the hair is attained.

I have been offering samples of mountain wool to my friends, demanding in return that the fineness of wildness be fairly recognized and confessed, but the returns are deplorably tame. The first question asked is, "Now truly, wild sheep, wild sheep, have you any wool?" while they peer curiously down among the hairs through lenses and spectacles. "Yes, wild sheep, you *have* wool; but Mary's lamb had more. In the name of use, how many wild sheep, think you, would be required to furnish wool sufficient for a pair of socks?" I endeavor to point out the irrelevancy of the latter question, arguing that wild wool was not made for

man but for sheep, and that, however deficient
as clothing for other animals, it is just the thing
for the brave mountain-dweller that wears it.
Plain, however, as all this appears, the quantity question rises again and again in all its
commonplace tameness. For in my experience
it seems well-nigh impossible to obtain a hearing on behalf of Nature from any other standpoint than that of human use. Domestic flocks
yield more flannel per sheep than the wild,
therefore it is claimed that culture has improved upon wildness; and so it has as far as
flannel is concerned, but all to the contrary as
far as a sheep's dress is concerned. If every
wild sheep inhabiting the Sierra were to put
on tame wool, probably only a few would sur-
vive the dangers of a single season. With their
fine limbs muffled and buried beneath a tangle
of hairless wool, they would become short-
winded, and fall an easy prey to the strong
mountain wolves. In descending precipices
they would be thrown out of balance and
killed, by their taggy wool catching upon
sharp points of rocks. Disease would also be
brought on by the dirt which always finds a
lodgment in tame wool, and by the draggled
and water-soaked condition into which it falls
during stormy weather.

No dogma taught by the present civilization

seems to form so insuperable an obstacle in the way of a right understanding of the relations which culture sustains to wildness as that which regards the world as made especially for the uses of man. Every animal, plant, and crystal controverts it in the plainest terms. Yet it is taught from century to century as something ever new and precious, and in the resulting darkness the enormous conceit is allowed to go unchallenged.

I have never yet happened upon a trace of evidence that seemed to show that any one animal was ever made for another as much as it was made for itself. Not that Nature manifests any such thing as selfish isolation. In the making of every animal the presence of every other animal has been recognized. Indeed, every atom in creation may be said to be acquainted with and married to every other, but with universal union there is a division sufficient in degree for the purposes of the most intense individuality; no matter, therefore, what may be the note which any creature forms in the song of existence, it is made first for itself, then more and more remotely for all the world and worlds.

Were it not for the exercise of individualizing cares on the part of Nature, the universe would be felted together like a fleece of tame wool.

WILD WOOL

But we are governed more than we know, and most when we are wildest. Plants, animals, and stars are all kept in place, bridled along appointed ways, *with* one another, and *through the midst* of one another — killing and being killed, eating and being eaten, in harmonious proportions and quantities. And it is right that we should thus reciprocally make use of one another, rob, cook, and consume, to the utmost of our healthy abilities and desires. Stars attract one another as they are able, and harmony results. Wild lambs eat as many wild flowers as they can find or desire, and men and wolves eat the lambs to just the same extent.

This consumption of one another in its various modifications is a kind of culture varying with the degree of directness with which it is carried out, but we should be careful not to ascribe to such culture any improving qualities upon those on whom it is brought to bear. The water-ouzel plucks moss from the river-bank to build its nest, but it does not improve the moss by plucking it. We pluck feathers from birds, and less directly wool from wild sheep, for the manufacture of clothing and cradle-nests, without improving the wool for the sheep, or the feathers for the bird that wore them. When a hawk pounces upon a linnet and proceeds to pull out its feathers, preparatory to making a meal, the hawk may be said to be cultivating the linnet, and he certainly does effect an improvement as far as hawk-food is concerned; but what of the songster? He ceases to be a linnet as soon as he is snatched from the woodland choir; and when, hawklike, we snatch the wild sheep from its native rock, and, instead of eating and wearing it at once, carry it home, and breed the hair out of its wool and the bones out of its body, it ceases to be a sheep.

These breeding and plucking processes are similarly improving as regards the secondary uses aimed at; and, although the one requires but a few minutes for its accomplishment, the other many years or centuries, they are essentially alike. We eat wild oysters alive with great directness, waiting for no cultivation, and leaving scarce a second of distance between the shell and the lip; but we take wild sheep home and subject them to the many extended processes of husbandry, and finish by boiling them in a pot — a process which completes all sheep improvements as far as man is concerned. It will be seen, therefore, that wild wool and tame wool — wild sheep and tame sheep — are terms not properly comparable, nor are they in any correct sense to be considered as bearing any antagonism toward each other; they are different things, planned and accomplished for wholly different purposes.

Illustrative examples bearing upon this interesting subject may be multiplied indefinitely, for they abound everywhere in the plant and animal kingdoms wherever culture has reached. Recurring for a moment to apples. The beauty and completeness of a wild apple tree living its own life in the woods is heartily acknowledged by all those who have been so happy as to form its acquaintance. The fine wild piquancy of its fruit is unrivaled, but in the great question of quantity as human food wild apples are found wanting. Man, therefore, takes the tree from the woods, manures and prunes and grafts, plans and guesses, adds a little of this and that, selects and rejects, until apples of every conceivable size and softness are produced, like nut-galls in response to the irritating punctures of insects. Orchard apples are to me the most eloquent words that culture has ever spoken, but they reflect no imperfection upon Nature's spicy crab. Every cultivated apple is a crab, not improved, *but cooked*, variously softened and swelled out in the process, mellowed, sweetened, spiced, and rendered pulpy and foodful, but as utterly unfit for the uses of nature as a meadowlark killed and plucked and roasted. Give to Nature every cultured apple — codling, pippin, russet — and every sheep so laboriously compounded — muffled Southdowns, hairy Cotswolds, wrinkled Merinos — and she would throw the one to her caterpillars, the other to her wolves.

It is now some thirty-six hundred years since Jacob kissed his mother and set out across the plains of Padan-aram to begin his experiments upon the flocks of his uncle, Laban; and, notwithstanding the high degree of excellence he attained as a wool-grower, and the innumerable painstaking efforts subsequently made by individuals and associations in all kinds of pastures and climates, we still seem to be as far from definite and satisfactory results as we ever were. In one breed the wool is apt to wither and crinkle like hay on a sunbeaten hillside. In another, it is lodged and matted together like the lush tangled grass of a manured meadow. In one the staple is deficient in length, in another in fineness; while in all there is a constant tendency toward disease, rendering various washings and dippings indispensable to prevent its falling out. The problem of the quality and quantity of the carcass seems to be as doubtful and as far removed from a satisfactory solution as that of the wool.

WILD WOOL

Desirable breeds blundered upon by long series of groping experiments are often found to be unstable and subject to disease — bots, foot-rot, blind-staggers, etc. — causing infinite trouble, both among breeders and manufacturers. Would it not be well, therefore, for some one to go back as far as possible and take a fresh start?

The source or sources whence the various breeds were derived is not positively known, but there can be hardly any doubt of their being descendants of the four or five wild species so generally distributed throughout the mountainous portions of the globe, the marked differences between the wild and domestic species being readily accounted for by the known variability of the animal, and by the long series of painstaking selection to which all its characteristics have been subjected. No other animal seems to yield so submissively to the manipulations of culture. Jacob controlled the color of his flocks merely by causing them to stare at objects of the desired hue; and possibly Merinos may have caught their wrinkles from the perplexed brows of their breeders. The California species (*Ovis montana*)[1] is a

[1] The wild sheep of California are now classified as *Ovis nelsoni*. Whether those of the Shasta region belonged to the latter species, or to the bighorn species of Oregon, Idaho, and Washington, is still an unsettled question. [Editor.]

noble animal, weighing when full-grown some three hundred and fifty pounds, and is well worthy the attention of wool-growers as a point from which to make a new departure. That it will breed with the domestic sheep I have not the slightest doubt, and I cordially recommend the experiment to the various wool-growers' associations as one of great national importance. From my knowledge of the homes and habits of our wild sheep I feel confident that several hundred could be obtained for breeding purposes from the Sierra alone, and I am ready to undertake their capture. A little pure wildness is the one great present want, both of men and sheep.

LIKE the forests of Washington, already
described, those of Oregon are in great part
made up of the Douglas spruce,[1] or Oregon
pine (*Abies Douglasii*). A large number of
mills are at work upon this species, especially
along the Columbia, but these as yet have
made but little impression upon its dense
masses, the mills here being small as compared
with those of the Puget Sound region. The
white cedar, or Port Orford cedar (*Cupressus
Lawsoniana*, or *Chamæcyparis Lawsoniana*), is
one of the most beautiful of the evergreens, and
produces excellent lumber, considerable quantities of which are shipped to the San Francisco market. It is found mostly about Coos
Bay, along the Coquille River, and on the
northern slopes of the Siskiyou Mountains,
and extends down the coast into California.
The silver firs, the spruces, and the colossal
arbor-vitæ, or white cedar [2] (*Thuja gigantea*),
described in the chapter on Washington, are

---

[1] *Pseudotsuga taxifolia*. Brit. [Editor.]
[2] *Thuja plicata* Don. [Editor.]

also found here in great beauty and perfection, the largest of these (*Picea grandis*, Loud.; *Abies grandis*, Lindl.) being confined mostly to the coast region, where it attains a height of three hundred feet, and a diameter of ten or twelve feet. Five or six species of pines are found in the State, the most important of which, both as to lumber and as to the part they play in the general wealth and beauty of the forests, are the yellow and sugar pines (*Pinus ponderosa* and *P. Lambertiana*). The yellow pine is most abundant on the eastern slopes of the Cascades, forming there the main bulk of the forest in many places. It is also common along the borders of the open spaces in Willamette Valley. In the southern portion of the State the sugar pine, which is the king of all the pines and the glory of the Sierra forests, occurs in considerable abundance in the basins of the Umpqua and Rogue Rivers, and it was in the Umpqua Hills that this noble tree was first discovered by the enthusiastic botanical explorer David Douglas, in the year 1826.

This is the Douglas for whom the noble Douglas spruce is named, and many a fair blooming plant also, which will serve to keep his memory fresh and sweet as long as beautiful trees and flowers are loved. The Indians

of the lower Columbia River watched him
with lively curiosity as he wandered about in
the woods day after day, gazing intently on
the ground or at the great trees, collecting
specimens of everything he saw, but, unlike
all the eager fur-gathering strangers they had
hitherto seen, caring nothing about trade. And
when at length they came to know him better,
and saw that from year to year the growing
things of the woods and prairies, meadows
and plains, were his only object of pursuit,
they called him the "Man of Grass," a title
of which he was proud.

He was a Scotchman and first came to this
coast in the spring of 1825 under the auspices
of the London Horticultural Society, landing
at the mouth of the Columbia after a long,
dismal voyage of eight months and fourteen
days. During this first season he chose Fort
Vancouver, belonging to the Hudson's Bay
Company, as his headquarters, and from there
made excursions into the glorious wilderness
in every direction, discovering many new
species among the trees as well as among
the rich underbrush and smaller herbaceous
vegetation. It was while making a trip to
Mount Hood this year that he discovered the
two largest and most beautiful firs in the
world (*Picea amabilis* and *P. nobilis* — now

called *Abies*), and from the seeds which he then collected and sent home tall trees are now growing in Scotland.

In one of his trips that summer, in the lower Willamette Valley, he saw in an Indian's tobacco-pouch some of the seeds and scales of a new species of pine, which he learned were gathered from a large tree that grew far to the southward. Most of the following season was spent on the upper waters of the Columbia, and it was not until September that he returned to Fort Vancouver, about the time of the setting-in of the winter rains. Nevertheless, bearing in mind the great pine he had heard of, and the seeds of which he had seen, he made haste to set out on an excursion to the headwaters of the Willamette in search of it; and how he fared on this excursion and what dangers and hardships he endured is best told in his own journal, part of which I quote as follows: —

October 26th, 1826. Weather dull. Cold and cloudy. When my friends in England are made acquainted with my travels I fear they will think that I have told them nothing but my miseries. . . . I quitted my camp early in the morning to survey the neighboring country, leaving my guide to take charge of the horses until my return in the evening. About an hour's walk from the camp I met an Indian, who on perceiving me instantly strung his

bow, placed on his left arm a sleeve of raccoon skin
and stood on the defensive. Being quite sure that
conduct was prompted by fear and not by hostile
intentions, the poor fellow having probably never
seen such a being as myself before, I laid my gun
at my feet on the ground and waved my hand for
him to come to me, which he did slowly and with
great caution. I then made him place his bow and
quiver of arrows beside my gun, and striking a
light gave him a smoke out of my own pipe and
a present of a few beads. With my pencil I made
a rough sketch of the cone and pine tree which I
wanted to obtain and drew his attention to it, when
he instantly pointed with his hand to the hills
fifteen or twenty miles distant towards the south;
and when I expressed my intention of going thither,
cheerfully set about accompanying me. At midday
I reached my long-wished-for pines and lost no
time in examining them and endeavoring to collect
specimens and seeds. New and strange things seldom fail to make strong impressions and are therefore frequently overrated; so that, lest I should
never see my friends in England to inform them
verbally of this most beautiful and immensely
grand tree, I shall here state the dimensions of the
largest I could find among several that had been
blown down by the wind. At three feet from the
ground its circumference is fifty-seven feet, nine
inches; at one hundred and thirty-four feet, seven-
teen feet five inches; the extreme length two hundred and forty-five feet. . . . As it was impossible
either to climb the tree or hew it down, I endeavored
to knock off the cones by firing at them with ball,
when the report of my gun brought eight Indians,

all of them painted with red earth, armed with bows, arrows, bone-tipped spears, and flint knives. They appeared anything but friendly. I explained to them what I wanted and they seemed satisfied and sat down to smoke; but presently I saw one of them string his bow and another sharpen his flint knife with a pair of wooden pincers and suspend it on the wrist of his right hand. Further testimony of their intentions was unnecessary. To save myself by flight was impossible, so without hesitation I stepped back about five paces, cocked my gun, drew one of the pistols out of my belt, and holding it in my left hand, the gun in my right, showed myself determined to fight for my life. As much as possible I endeavored to preserve my coolness, and thus we stood looking at one another without making any movement or uttering a word for perhaps ten minutes, when one at last, who seemed to be the leader, gave a sign that they wished for some tobacco; this I signified they should have if they fetched a quantity of cones. They went off immediately in search of them, and no sooner were they all out of sight than I picked up my three cones and some twigs of the trees and made the quickest possible retreat, hurrying back to my camp, which I reached before dusk. The Indian who last undertook to be my guide to the trees I sent off before gaining my encampment, lest he should betray me. How irksome is the darkness of night to one under such circumstances. I cannot speak a word to my guide, nor have I a book to divert my thoughts, which are continually occupied with the dread lest the hostile Indians should trace me hither and make an attack. I now write lying on the grass with my gun cocked beside me, and penning these lines by the light of my *Columbian candle*, namely, an ignited piece of rosin-wood.

Douglas named this magnificent species *Pinus Lambertiana*, in honor of his friend Dr. Lambert, of London. This is the noblest pine thus far discovered in the forests of the world, surpassing all others not only in size but in beauty and majesty. Oregon may well be proud that its discovery was made within her borders, and that, though it is far more abundant in California, she has the largest known specimens. In the Sierra the finest sugar pine forests lie at an elevation of about five thousand feet. In Oregon they occupy much lower ground, some of the trees being found but little above tide-water.

No lover of trees will ever forget his first meeting with the sugar pine. In most coniferous trees there is a sameness of form and expression which at length becomes wearisome to most people who travel far in the woods. But the sugar pines are as free from conventional forms as any of the oaks. No two are so much alike as to hide their individuality from any observer. Every tree is appreciated as a study in itself and proclaims in no uncertain terms the surpassing grandeur of the species. The branches, mostly near the summit, are sometimes nearly forty feet long, feathered richly all around with short, leafy branchlets, and tasselled with cones a foot and a half long. And when these superb arms are outspread, radiating in every direction, an immense crown-like mass is formed which, poised on the noble shaft and filled with sunshine, is one of the grandest forest objects conceivable. But though so wild and unconventional when full-grown, the sugar pine is a remarkably regular tree in youth, a strict follower of coniferous fashions, slim, erect, tapering, symmetrical, every branch in place. At the age of fifty or sixty years this shy, fashionable form begins to give way. Special branches are thrust out away from the general outlines of the trees and bent down with cones. Henceforth it becomes more and more original and independent in style, pushes boldly aloft into the winds and sunshine, growing ever more stately and beautiful, a joy and inspiration to every beholder.

Unfortunately, the sugar pine makes excellent lumber. It is too good to live, and is already passing rapidly away before the woodman's axe. Surely out of all of the abounding forest-wealth of Oregon a few specimens might be spared to the world, not as dead lumber, but as living trees. A park of moderate

THE FORESTS OF OREGON

extent might be set apart and protected for
public use forever, containing at least a few
hundreds of each of these noble pines, spruces,
and firs. Happy will be the men who, having
the power and the love and benevolent forecast
to do this, will do it. They will not be forgotten. The trees and their lovers will sing their
praises, and generations yet unborn will rise
up and call them blessed.

Dotting the prairies and fringing the edges
of the great evergreen forests we find a considerable number of hardwood trees, such as
the oak, maple, ash, alder, laurel, madrone,
flowering dogwood, wild cherry, and wild
apple. The white oak (*Quercus Garryana*) is
the most important of the Oregon oaks as a
timber tree, but not nearly so beautiful as
Kellogg's oak (*Q. Kelloggii*). The former is
found mostly along the Columbia River, particularly about the Dalles, and a considerable quantity of useful lumber is made from
it and sold, sometimes for eastern white oak,
to wagon-makers. Kellogg's oak is a magnificent tree and does much for the picturesque
beauty of the Umpqua and Rogue River Valleys where it abounds. It is also found in all
the Yosemite valleys of the Sierra, and its
acorns form an important part of the food
of the Digger Indians. In the Siskiyou Moun-

251

tains there is a live oak (*Q. chrysolepis*), wide-spreading and very picturesque in form, but not very common. It extends southward along the western flank of the Sierra and is there more abundant and much larger than in Oregon, oftentimes five to eight feet in diameter.

The maples are the same as those in Washington, already described, but I have not seen any maple groves here equal in extent or in the size of the trees to those on the Snoqualmie River.

The Oregon ash is now rare along the streambanks of western Oregon, and it grows to a good size and furnishes lumber that is for some purposes equal to the white ash of the Western States.

Nuttall's flowering dogwood makes a brave display with its wealth of showy involucres in the spring along cool streams. Specimens of the flowers may be found measuring eight inches in diameter.

The wild cherry (*Prunus emarginata*, var. *mollis*) is a small, handsome tree seldom more than a foot in diameter at the base. It makes valuable lumber and its black, astringent fruit furnishes a rich resource as food for the birds. A smaller form is common in the Sierra, the fruit of which is eagerly eaten by the Indians and hunters in time of need.

THE FORESTS OF OREGON

The wild apple (*Pyrus rivularis*) is a fine, hearty, handsome little tree that grows well in rich, cool soil along streams and on the edges of beaver-meadows from California through Oregon and Washington to southeastern Alaska. In Oregon it forms dense, tangled thickets, some of them almost impenetrable. The largest trunks are nearly a foot in diameter. When in bloom it makes a fine show with its abundant clusters of flowers, which are white and fragrant. The fruit is very small and savagely acid. It is wholesome, however, and is eaten by birds, bears, Indians, and many other adventurers, great and small.

Passing from beneath the shadows of the woods where the trees grow close and high, we step into charming wild gardens full of lilies, orchids, heathworts, roses, etc., with colors so gay and forming such sumptuous masses of bloom, they make the gardens of civilization, however lovingly cared for, seem pathetic and silly. Around the great fire-mountains, above the forests and beneath the snow, there is a flowery zone of marvelous beauty planted with anemones, erythroniums, daisies, bryanthus, kalmia, vaccinium, cassiope, saxifrages, etc., forming one continuous garden fifty or sixty miles in circumference, and so deep and luxuriant and closely woven it seems as if Nature, glad to find an opening, were economizing space and trying to see how many of her bright-eyed darlings she can get together in one mountain wreath.

Along the slopes of the Cascades, where the woods are less dense, especially about the headwaters of the Willamette, there are miles of rhododendron, making glorious outbursts of purple bloom, and down on the prairies in rich, damp hollows the blue-flowered camassia grows in such profusion that at a little distance its dense masses appear as beautiful blue lakes imbedded in the green, flowery plains; while all about the streams and the lakes and the beaver-meadows and the margins of the deep woods there is a magnificent tangle of gaultheria and huckleberry bushes with their myriads of pink bells, reinforced with hazel, cornel, rubus of many species, wild plum, cherry, and crab apple; besides thousands of charming bloomers to be found in all sorts of places throughout the wilderness whose mere names are refreshing, such as linnæa, menziesia, pyrola, chimaphila, brodiæa, smilacina, fritillaria, calochortus, trillium, clintonia, veratrum, cypripedium, goodyera, spiranthes, habenaria, and the rare and lovely "Hider of the North," *Calypso borealis*, to find which is alone a sufficient object for a journey into the wilderness. And besides these there is a charming underworld of ferns and mosses flourishing gloriously beneath all the woods.

Everybody loves wild woods and flowers more or less. Seeds of all these Oregon evergreens and of many of the flowering shrubs and plants have been sent to almost every country under the sun, and they are now growing in carefully tended parks and gardens. And now that the ways of approach are open one would expect to find these woods and gardens full of admiring visitors reveling in their beauty like bees in a clover-field. Yet few care to visit them. A portion of the bark of one of the California trees, the mere dead skin, excited the wondering attention of thousands when it was set up in the Crystal Palace in London, as did also a few peeled spars, the shafts of mere saplings from Oregon or Washington. Could one of these great silver firs or sugar pines three hundred feet high have been transplanted entire to that exhibition, how enthusiastic would have been the praises accorded to it!

Nevertheless, the countless hosts waving at home beneath their own sky, beside their own noble rivers and mountains, and standing on a flower-enameled carpet of mosses thousands of square miles in extent, attract but little attention. Most travelers content themselves with what they may chance to see from car windows, hotel verandas, or the deck of a steamer on the lower Columbia — clinging to the battered highways like drowning sailors to a life-raft. When an excursion into the woods is proposed, all sorts of exaggerated or imaginary dangers are conjured up, filling the kindly, soothing wilderness with colds, fevers, Indians, bears, snakes, bugs, impassable rivers, and jungles of brush, to which is always added quick and sure starvation.

As to starvation, the woods are full of food, and a supply of bread may easily be carried for habit's sake, and replenished now and then at outlying farms and camps. The Indians are seldom found in the woods, being confined mainly to the banks of the rivers, where the greater part of their food is obtained. Moreover, the most of them have been either buried since the settlement of the country or civilized into comparative innocence, industry, or harmless laziness. There are bears in the woods, but not in such numbers nor of such unspeakable ferocity as town-dwellers imagine, nor do bears spend their lives in going about the country like the devil, seeking whom they may devour. Oregon bears, like most others, have no liking for man either as meat or as society; and while some may be curious at times to see what manner of creature he is, most of them have learned to shun people as deadly enemies. They have been poisoned, trapped, and shot at until they have become shy, and it is no longer easy to make their acquaintance. Indeed, since the settlement of the country, notwithstanding far the greater portion is yet wild, it is difficult to find any of the larger animals that once were numerous and comparatively familiar, such as the bear, wolf, panther, lynx, deer, elk, and antelope.

As early as 1843, while the settlers numbered only a few thousands, and before any sort of government had been organized, they came together and held what they called "a wolf meeting," at which a committee was appointed to devise means for the destruction of wild animals destructive to tame ones, which committee in due time begged to report as follows: —

It being admitted by all that bears, wolves, panthers, etc., are destructive to the useful animals owned by the settlers of this colony, your committee would submit the following resolutions as the sense of this meeting, by which the community may be governed in carrying on a defensive and destructive war on all such animals: —

Resolved, 1st. — That we deem it expedient for the community to take immediate measures for the destruction of all wolves, panthers and bears, and such other animals as are known to be destructive to cattle, horses, sheep and hogs.

2d. — That a bounty of fifty cents be paid for the destruction of a small wolf, $3.00 for a large wolf, $1.50 for a lynx, $2.00 for a bear and $5.00 for a panther.

This center of destruction was in the Willamette Valley. But for many years prior to the beginning of the operations of the "Wolf Organization" the Hudson's Bay Company had established forts and trading-stations over all the country, wherever fur-gathering Indians could be found, and vast numbers of these animals were killed. Their destruction has since gone on at an accelerated rate from year to year as the settlements have been extended, so that in some cases it is difficult to obtain specimens enough for the use of naturalists. But even before any of these settlements were made, and before the coming of the Hudson's Bay Company, there was very little danger to be met in passing through this wilderness as far as animals were concerned, and but little of any kind as compared with the dangers encountered in crowded houses and streets.

When Lewis and Clark made their famous

THE FORESTS OF OREGON

trip across the continent in 1804–05, when all the Rocky Mountain region was wild, as well as the Pacific Slope, they did not lose a single man by wild animals, nor, though frequently attacked, especially by the grizzlies of the Rocky Mountains, were any of them wounded seriously. Captain Clark was bitten on the hand by a wolf as he lay asleep; that was one bite among more than a hundred men while traveling through eight to nine thousand miles of savage wilderness. They could hardly have been so fortunate had they stayed at home. They wintered on the edge of the Clatsop plains, on the south side of the Columbia River near its mouth. In the woods on that side they found game abundant, especially elk, and with the aid of the friendly Indians who furnished salmon and "wapatoo" (the tubers of *Sagittaria variabilis*), they were in no danger of starving.

But on the return trip in the spring they reached the base of the Rocky Mountains when the range was yet too heavily snow-laden to be crossed with horses. Therefore they had to wait some weeks. This was at the head of one of the northern branches of Snake River, and, their scanty stock of provisions being nearly exhausted, the whole party was compelled to live mostly on bears and dogs; deer, antelope, and elk, usually abundant, were now scarce because the region had been closely hunted over by the Indians before their arrival.

Lewis and Clark had killed a number of bears and saved the skins of the more interesting specimens, and the variations they found in size, color of the hair, etc., made great difficulty in classification. Wishing to get the opinion of the Chopumish Indians, near one of whose villages they were encamped, concerning the various species, the explorers unpacked their bundles and spread out for examination all the skins they had taken. The Indian hunters immediately classed the white, the deep and the pale grizzly red, the grizzly dark-brown — in short, all those with the extremities of the hair of a white or frosty color without regard to the color of the ground or foil — under the name of *hoh-host*. The Indians assured them that these were all of the same species as the white bear, that they associated together, had longer nails than the others, and never climbed trees. On the other hand, the black skins, those that were black with white hairs intermixed or with a white breast, the uniform bay, the brown, and the light reddish-brown, were classed under the name *yack-ah*, and were said to resemble each other in being smaller and having shorter nails, in climbing trees, and being so little vicious that they could be pursued with safety.

Lewis and Clark came to the conclusion that all those with white-tipped hair found by them in the basin of the Columbia belonged to the same species as the grizzlies of the upper Missouri; and that the black and reddish-brown, etc., of the Rocky Mountains belong to a second species equally distinct from the grizzly and the black bear of the Pacific Coast and the East, which never vary in color.

As much as possible should be made by the ordinary traveler of these descriptions, for he will be likely to see very little of any species for himself; not that bears no longer exist here, but because, being shy, they keep out of the way. In order to see them and learn their habits one must go softly and alone, lingering long in the fringing woods on the banks of the salmon streams, and in the small openings in the midst of thickets where berries are most abundant.

As for rattlesnakes, the other grand dread of town-dwellers when they leave beaten roads, there are two, or perhaps three, species of them in Oregon. But they are nowhere to be found in great numbers. In western Oregon they are hardly known at all. In all my walks in the

Oregon forest I have never met a single specimen, though a few have been seen at long
intervals.

When the country was first settled by the
whites, fifty years ago, the elk roamed through
the woods and over the plains to the east of
the Cascades in immense numbers; now
they are rarely seen except by experienced
hunters who know their haunts in the deepest
and most inaccessible solitudes to which they
have been driven. So majestic an animal
forms a tempting mark for the sportsman's
rifle. Countless thousands have been killed
for mere amusement and they already seem
to be nearing extinction as rapidly as the
buffalo. The antelope also is vanishing from
the Columbia plains before the farmers and
cattle-men. Whether the moose still lingers
in Oregon or Washington I am unable to say.

On the highest mountains of the Cascade
Range the wild goat roams in comparative
security, few of his enemies caring to go so far
in pursuit and to hunt on ground so high and
so dangerous. He is a brave, sturdy, shaggy
mountaineer of an animal, enjoying the freedom and security of crumbling ridges and
overhanging cliffs above the glaciers, oftentimes beyond the reach of the most daring
hunter. They seem to be as much at home on

the ice and snow-fields as on the crags, making their way in flocks from ridge to ridge on the great volcanic mountains by crossing the glaciers that lie between them, traveling in single file guided by an old experienced leader, like a party of climbers on the Alps. On these ice-journeys they pick their way through networks of crevasses and over bridges of snow with admirable skill, and the mountaineer may seldom do better in such places than to follow their trail, if he can. In the rich alpine gardens and meadows they find abundance of food, venturing sometimes well down in the prairie openings on the edge of the timber-line, but holding themselves ever alert and watchful, ready to flee to their highland castles at the faintest alarm. When their summer pastures are buried beneath the winter snows, they make haste to the lower ridges, seeking the wind-beaten crags and slopes where the snow cannot lie at any great depth, feeding at times on the leaves and twigs of bushes when grass is beyond reach.

The wild sheep is another admirable alpine rover, but comparatively rare in the Oregon mountains, choosing rather the drier ridges to the southward on the Cascades and to the eastward among the spurs of the Rocky Mountain chain.

Deer give beautiful animation to the forests, harmonizing finely in their color and movements with the gray and brown shafts of the trees and the swaying of the branches as they stand in groups at rest, or move gracefully and noiselessly over the mossy ground about the edges of beaver-meadows and flowery glades, daintily culling the leaves and tips of the mints and aromatic bushes on which they feed. There are three species, the black-tailed, white-tailed, and mule deer; the last being restricted in its range to the open woods and plains to the eastward of the Cascades. They are nowhere very numerous now, killing for food, for hides, or for mere wanton sport, having well-nigh exterminated them in the more accessible regions, while elsewhere they are too often at the mercy of the wolves.

Gliding about in their shady forest homes, keeping well out of sight, there is a multitude of sleek fur-clad animals living and enjoying their clean, beautiful lives. How beautiful and interesting they are is about as difficult for busy mortals to find out as if their homes were beyond sight in the sky.